THE ANALYTIC HIERARCHY PROCESS IN NATURAL RESOURCE AND ENVIRONMENTAL DECISION MAKING

Managing Forest Ecosystems

Volume 3

Series Editors:

Klaus von Gadow

Georg-August-University,
Göttingen, Germany

Timo Pukkala

University of Joensuu,
Joensuu, Finland

and

Margarida Tomé

Instituto Superior de Agronomía,
Lisbon, Portugal

Aims & Scope:

Well-managed forests and woodlands are a renewable resource, producing essential raw material with minimum waste and energy use. Rich in habitat and species diversity, forests may contribute to increased ecosystem stability. They can absorb the effects of unwanted deposition and other disturbances and protect neighbouring ecosystems by maintaining stable nutrient and energy cycles and by preventing soil degradation and erosion. They provide much-needed recreation and their continued existence contributes to stabilizing rural communities.

Forests are managed for timber production and species, habitat and process conservation. A subtle shift from *multiple-use management* to *ecosystems management* is being observed and the new ecological perspective of *multi-functional forest management* is based on the principles of ecosystem diversity, stability and elasticity, and the dynamic equilibrium of primary and secondary production.

Making full use of new technology is one of the challenges facing forest management today. Resource information must be obtained with a limited budget. This requires better timing of resource assessment activities and improved use of multiple data sources. Sound ecosystems management, like any other management activity, relies on effective forecasting and operational control.

The aim of the book series ***Managing Forest Ecosystems*** is to present state-of-the-art research results relating to the practice of forest management. Contributions are solicited from prominent authors. Each reference book, monograph or proceedings volume will be focused to deal with a specific context. Typical issues of the series are: resource assessment techniques, evaluating sustainability for even-aged and uneven-aged forests, multi-objective management, predicting forest development, optimizing forest management, biodiversity management and monitoring, risk assessment and economic analysis.

The titles published in this series are listed at the end of this volume.

The Analytic Hierarchy Process in Natural Resource and Environmental Decision Making

Edited by

Daniel L. Schmoldt

USDA Forest service, Southern Research station,
Blacksburg, VA, U.S.A.

Jyrki Kangas

Finnish Forest Research Institute,
Kannus Research Station, Kannus, Finland

Guillermo A. Mendoza

University of Illinois,
Department of Natural Resources and Environmental Sciences,
Urbana, IL, U.S.A.

and

Mauno Pesonen

Finnish Forest Research Institute,
Vantaa, Finland

KLUWER ACADEMIC PUBLISHERS

DORDRECHT / BOSTON / LONDON

A C.I.P. Catalogue record for this book is available from the Library of Congress.

ISBN 0-7923-7076-7

Published by Kluwer Academic Publishers,
P.O. Box 17, 3300 AA Dordrecht, The Netherlands.

Sold and distributed in North, Central and South America
by Kluwer Academic Publishers,
101 Philip Drive, Norwell, MA 02061, U.S.A.

In all other countries, sold and distributed
by Kluwer Academic Publishers,
P.O. Box 322, 3300 AH Dordrecht, The Netherlands.

Printed on acid-free paper

Printed in the Netherlands.

Contents

Contributors

Jyrki Ahola
Lappeenranta University of Technology, Department of Business Administration, Finland

Juha M. Alho
University of Joensuu, Department of Statistics, Joensuu Finland

Reza Banai
Graduate Program in City and Regional Planning, University of Memphis, Memphis TN USA

Luis Díaz-Balteiro
Departamento de Ingeniería Agrícola y Forestal. E.T.S. Ingenierías Agrarias. Avda. Madrid, Palencia, Spain

Kathleen Hirst
GIS Applications Pty Ltd., South Caulfield, Victoria, Australia

Leena A. Hytönen
Finnish Forest Research Institute, Kannus Research Station, Kannus, Finland

Robert M. Itami
Centre for GIS and Modelling, Department of Geomatics, University of Melbourne, Victoria, Australia

Piotr Jankowski
Department of Geography, University of Idaho, Moscow, ID USA

Miika Kajanus
Pohjois-Savo Polytechnic, Rural Education, Peltosalmi, Finland

Annika S. Kangas
Finnish Forest Research Institute, Kannus, Finland

Jyrki Kangas
Finnish Forest Research Institute, Kannus, Finland

Osmo Kolehmainen
University of Joensuu, Department of Statistics, Joensuu Finland

Pekka Korhonen
Helsinki School of Economics and Business Administration, Runeberginkatu, Helsinki, Finland

Mikko Kurttila
University of Joensuu, Faculty of Forestry, Finland

Pekka Leskinen
Finnish Forest Research Institute, Kannus Research Station, Kannus Finland

Teppo Loikkanen
Finnish Forest and Park Service, Finland

Glen MacLaren
Centre for GIS and Modelling, Department of Geomatics, University of Melbourne, Victoria, Australia

Guillermo A. Mendoza
Department of Natural Resources and Environmental Sciences, University of Illinois, Urbana, Illinois USA

Timothy Nyerges
Department of Geography, University of Washington, Seattle, WA USA

Mauno Pesonen
Finnish Forest Research Institute, Finland

David L. Peterson
USGS Forest and Rangeland Ecosystem Science Center, Cascadia Field Station, Seattle WA USA

R. Prabhu
Center for International Forestry Research (CIFOR) Regional Office, Institute of Environmental Studies, University of Zimbabwe, Harare, Zimbabwe

Timo Pukkala
University of Joensuu, Joensuu, Finland

Keith M. Reynolds
USDA Forest Service, Pacific Northwest Research Station, Corvallis, OR USA

Carlos Romero
Departamento de Economía y Gestión Forestal, E.T.S. Ingenieros de Montes, Avenida Complutense s/n Madrid, Spain

Thomas L. Saaty
University of Pittsburgh, Pittsburgh PA USA

Daniel L. Schmoldt
USDA Forest Service, Southern Research Station, Madison WI USA

Jyrki Wallenius
Helsinki School of Economics and Business Administration, Runeberginkatu, Helsinki, Finland

Preface

While different areas of human endeavour vary greatly in the types of problems encountered, there are many common elements that also make them alike. In particular, human activities are driven by needs, whether basic, physical needs or more ethereal social/psychological needs. Because need satisfaction (comfort) is not constant over time, people develop goals and objectives for their daily activities to provide a directed pathway to some desired comfort state—desire future condition (DFC). People must select their goals, their DFCs, and the activities that enable them to reach their DFCs. Because fiscal resources, natural resources, and human resources are finite, people must preferentially select certain goals, DFCs, and activities exclusive of others. These concepts are invariant whether one is talking about political policy, consumer purchasing, or managing lands with their component resources.

The primary aim for this book is to draw on the extensive body of research into, and applications of, the analytic hierarchy process (AHP), and to organize it into a single reference text for use by other scientists and students of decision making in the natural resource and environmental fields. Chapter contributions have been solicited from many researchers throughout the world, in addition to the chapters authored by the editors. These authored/solicited chapters cover applications of, as well as extensions to, the AHP in natural resources and the environment.

The book is organized into five sections followed by a final chapter which is intended to synthesize and summarize ideas scattered throughout the text. Each section is organized around a central theme related to the AHP. The first section is introductory in nature, wherein the first chapter discusses natural resource management in general and provides a straightforward example of how the AHP works in practice. Thomas Saaty graciously contributed chapter 2, which provides a rigorous treatment of the AHP's fundamentals. With these two chapters in hand, most readers should be able to digest the remaining chapters of the book, allowing for some difficulty with mathematical details in several cases.

Following these two introductory chapters, the second section covers integration of the AHP and mathematical optimisation. Chapters 3-6 are included here. Optimisation methods are very important decision-making tools because they provide the decision maker with a quantitative metric with which to choose the "best" alternative. Often, however, these methods require estimates of objective function parameters or selection of upper/lower bounds for constraints. Preference or likelihood values, derived from the AHP, can be used in the quantitative formulations of optimisation methods. Examples in these chapters include: multi-objective linear programming, heuristic optimisation, tactical planning using linear programming, and goal programming. In this way, the AHP enables mathematical programming techniques to include reliable estimates of important quantities, which are subjective and difficult to quantify.

The third section examines the use of the AHP as an aid to group decision making (chapters 7-10). Group decision making has become increasingly important for natural resource management and associated scientific applications. Multiple resource values must be treated coincidentally in time and space (multiple resource specialists included) and a large diversity of clientele must be included in decision processes (multiple stakeholders). The AHP is well suited to this type of decision scenario owing to its ability to readily incorporate multiple judgments. Applications of the AHP in this section include: fire research priority assessment in a workshop setting, prioritising criteria and indicators for sustainable forestry, natural resources planning, and mental models of spatial relationships. Furthermore, the AHP's natural hierarchical decomposition of a problem provides groups with a clear and understandable forum for deliberation.

There are a number of other decision methods—e.g. multi-attribute utility theory, SWOT analysis, and SMART—that can be combined with the AHP. Several chapters (11-13) describe applications that cover each of these alternative decision methods and that incorporate the AHP in some useful way. Examples cover: national timber harvest budgets, forest industry investment strategies, and prioritising watershed habitat restoration.

The fifth section of the text looks at some valuable extensions to the standard AHP as described by Saaty. In that section, authors describe how the AHP can be modified using the approximate reasoning of fuzzy sets, how its priority calculus can be modified, and how the AHP can incorporate spatial information or be part of an analysis of spatial information. Chapters 14-17 cover these topics.

The final chapter takes a broader view and considers some of the AHP's functionality in the context of natural resource problems, in general. This is done by examining those AHP features that contribute most to its value as a decision aide. Then, based on the AHP's functionality in its current form, we provide some insights into how the AHP might be extended further to make it even more valuable.

While this text could not include all applications of the AHP in the natural resources arena, we feel that we have provided a good overview of the method's potential. Furthermore, the set of referenced works from all chapters combined should serve as a comprehensive coverage of AHP applications in natural resources. As with any scientific review of this sort, it provides only a snapshot of current and past activities, but also offers a point of departure for new, innovative, and ambitious efforts by colleagues. We anxiously look forward to those future contributions and hope that formal decision processes, such as the AHP, can eventually become a regular part of land management decision making.

Daniel L. Schmoldt

Jyrki Kangas

Guillermo A. Mendoza

Mauno Pesonen

Foreword

Much like the business world that I am familiar with, managers of natural resources find themselves overwhelmed with data and analytical tools, but seriously lacking procedures and methods to integrate that information for sound, accountable decision making. Almost every day, one can pick up a newspaper and read about contentious environmental/ecological issues. Due to an ever-increasing human population, there are greater resource demands (increased consumption), multiple and often incompatible land uses are squeezed into closer proximity (conflicting use), and special interests have greater membership and political/economic clout (conflicting goals). As issues become more complex, there is an increasing need to apply more formal decision procedures. The analytic hierarchy process (AHP), which I developed in the 1970's, is such a method.

The AHP approach possesses three valuable characteristics that aid decision making. First, the AHP enables decision makers to structure a problem into a hierarchy consisting of a goal and subordinate features (decomposition). Second, pairwise comparisons between elements at each level enable a preferential ordering of decision elements (evaluation). Third, matrix algebra propagates level-specific, local priorities to global priorities (synthesis). Subordinate levels of the hierarchy, may include objectives, scenarios, events, actions, outcomes, and alternatives. Alternatives to be compared appear at the lowest level of the hierarchy. Because much natural resource decision making involves selecting among (or prioritising) a finite set of alternative courses of action, the AHP's characterization of decision making is particularly useful.

The several sections of this book cover a number of important areas related to application of the AHP. One of the important uses of the AHP in natural resource decision making is the integration of subjective preference with traditional, decision optimisation tools. Several of this book's chapters merge the AHP with mathematical programming analyses. Group decision making is another important area of application, as few workplace decisions—including those in natural resource—are made unilaterally. Decisions are typically made in consultation with others, or by a group, in a participatory environment. A third section of the text includes contributions that combine the AHP with other decision-making techniques (e.g., SMART), borrowing the advantages of both methods. A final section introduces methods that expand applications of the AHP to, for example, fuzzy logic, priority analysis, and spatially referenced data. In some chapters, the AHP is embedded into other software tools and decision processes, and in other chapters, real-world examples of its use are provided.

It is particularly satisfying for me to see my work take on some measure of importance in areas foreign to me, such as natural resources and the environment. I am continually amazed by the extent to which the AHP is applied outside of the business and economics arena. Much of my gratification comes from the realization that the AHP is truly a universal method for thinking and decision making. In addition, I am humbled and honoured that colleagues view my work with such high regard and continue to find new uses for it.

The editors of this book (and others in the natural resource community) have taken note of the AHP's potential to aid decision makers—in their thinking, in their decision analysis, and in decision accountability. The current text provides an excellent overview of past research in this area and illustrates current developments that further extend application of the AHP method in new directions. Given the importance that natural and environmental resources hold for all of us, I am delighted to see that members of that scientific community are using, adapting, and extending my work.

Thomas L. Saaty

Chapter 1

Basic Principles of Decision Making in Natural Resources and the Environment

Daniel L. Schmoldt
USDA Forest Service, Southern Research Station, Madison WI USA

Jyrki Kangas
Finnish Forest Research Institute, Kannus, Finland

G. A. Mendoza
Department of Natural Resources and Environmental Sciences, University of Illinois, Urbana, Illinois USA

Key words: Multiple objective decision making, decision analysis, preferences, natural resource management.

Abstract: As public land management merges biophysical, social, and economic objectives, management decision criteria become more extensive. Many of these criteria are value-laden, and yet are not easily expressed in monetary terms. Utility theory has traditionally been the decision model proffered by the management science and operations research communities. More recently, however, the analytic hierarchy process (AHP) has also received considerable attention, primarily because it places greater emphasis on the decision makers' preferences structures. A simple example of the AHP, for college enrolment, illustrates many of the method's salient features, and some of the underlying mathematics. A brief review of some applications of the AHP in natural resources management is also included. Land management agencies need to establish decision models that provide some structure for how decision-support information is organized and applied, so that decisions are made openly within a well-defined framework. In doing so, decision accountability and justification are achieved concomitantly with the process itself.

D.L. Schmoldt et al. (eds.),
The Analytic Hierarchy Process in Natural Resource and Environmental Decision Making, 1–13.

1. INTRODUCTION

Early in the beginning of the twentieth century, Gifford Pinchot defined effective natural resource management as "...providing the greatest good for the greatest number, in the long run" (q.v., Pinchot 1947). Interpretation of this principle has varied over time, depending on how the pivotal terms "greatest" and "good" were defined. Nevertheless, the statement's essence still retains validity as active land management enters the next century. Once satisfactory agreement is reached on those value judgments ("good" and "greatest"), it remains to determine *how*, *where*, *when*, *how much*, etc. Someone must decide among alternative courses of action, so that future events can achieve desired values.

The most recent embodiment of the Pinchot principle is *ecosystem management*. In this paradigm, an attempt has been made to remove the chasm that has treated people as separate from their biophysical environment (Unger and Salwasser 1991, FEMAT 1993, Lackey 1998). Now, biophysical, social/political/cultural, and economic processes together encompass the important interactions between people and the land resources upon which they depend. Economics has always been an important component of land management, but now social institutions (e.g., rural communities, indigenous cultures) and biophysical integrity (e.g., ecosystem processes, biodiversity) must also be considered. Nevertheless, one invariant that is still part of effective land management, regardless of how "effective" is defined, is the need to make rational and justifiable choices when faced with alternatives.

Multiplicity in land management objectives and in land management beneficiaries (which ultimately includes everyone) precludes the simultaneous satisfaction of everyone's wishes fully. In some cases, land management objectives are mutually inconsistent, and in other cases, our objectives cannot be fully met given practical limitations of space and time. Despite the fact that not everyone can have everything, land management must address the needs and desires of all stakeholders within the biophysical limitations of the land and the social and political institutions within which people live. To do so requires decision processes that are flexible and that are able to accommodate both subjective/qualitative and quantitative information.

Land management decision making is further handicapped by the uncertainty surrounding future events and by limitations of our knowledge about how the world works (Schmoldt and Rauscher 1996). Unforeseen changes in any of the biophysical, social/political, or economic components of management can render even the most "optimal" choice today ineffective tomorrow. Furthermore, the best science available can often only generalize about future scenarios because scientists do not thoroughly understand most ecosystem components individually, much less how they interact with each other and with human social and economic systems. As with multiple objectives and stakeholders, our decision methods must likewise address uncertainty and allow for periodic re-evaluation over time.

To aid human ability to understand and evaluate management situations and scenarios, a wide variety of analytical tools have been developed. These include, for example, simulation models, geographic information systems, expert systems, econometric models, and optimisation techniques under the umbrella of decision support (Reynolds *et al.* 1999). These aids are important adjuncts to good decision making, but each typically addresses only one aspect of land management. The decision maker must still integrate each tool's analytical results into a rational choice about *what* to do *where* and *when*. Decision analysis techniques take this natural next step to assist with selecting among competing alternatives. The following section provides a brief review of multiple criteria decision making and introduces the analytic hierarchy process (the subject of this book) as an important decision-making tool. After this decision analysis review, some of the existing literature and applications of the analytic hierarchy process (AHP) in natural resources are summarized.

2. MULTIPLE-OBJECTIVE DECISION MAKING

Given that people, monetary resources, facilities, equipment, time, and space are limited, most multi-objective decision problems cannot fully satisfy all objectives. Therefore, decision analysis attempts to compromise on some middle ground, covering all objectives, that maximizes "value" or "utility" or that minimizes "cost" or "loss"—where those terms are defined appropriately within the context of the problem at hand. Because, in most cases, the intent is to prescribe the best decision alternative (as opposed to describing how decisions are typically made, i.e. *behavioural analysis*), decision analysis is often referred to as *normative*. That is, a rational standard is prescribed as the best alternative, given the way that the current problem has been structured.

2.1 Normative Decision Making

The aim of any decision analysis is to lend support to decision making in problems that are too complex to be solved by the intuitive use of common sense alone. Strategic natural resource management decisions are typical examples of such problems. In a decision-theoretic approach, a decision is considered as a choice between two or more alternative measures. In a normative approach to decision-making, the starting point is that a rational decision-maker aims to choose the alternative which most probably maximizes the decision-maker's utility (or value system), based on information available to him or her on the decision alternatives (Kangas 1992). This is the viewpoint in the situation of a single decision maker. In group decision making, the total utility to be maximized can be taken as the combined utilities of the persons belonging to the group. In participatory decision-making processes, some or even all the decision-making power might be allocated to the participants.

In decision support, the aim is to ensure that the decision maker is as informed as possible. Information is produced regarding the decision situation, on alternative courses of action, and on consequences of alternative choices. A complete decision model constitutes the basis for decision support. Three things are included in the decision basis: the alternatives available, information about the consequences associated with these alternatives, and the preferences among these consequences (Bradshaw and Boose 1990). Keeney (1982) has divided decision analysis into *four* phases (the previous three plus one additional aspect): (1) structure the decision problem, (2) assess possible impacts of each alternative, (3) determine preferences of decision-makers, and (4) evaluate and compare decision alternatives. Each aspect of decision-support information has to be sound, so that the best, a good, or at least a satisfactory alternative can be selected. Errors or misinformation in any part of decision analysis can lead to questionable or invalid results.

In decision analysis, the decision situation is viewed holistically. Generally, numerical encoding of information concerning the decision situation can be taken as a precondition for an effective and thorough treatment of a complex decision problem (von Winterfeldt 1988, Guariso and Werthner 1989). Numerical decision analysis is based on logical axioms and a methodology founded on these axioms. This methodology must incorporate decision makers' and other stakeholders' preferences somehow.

A utility model is a mathematical tool that describes problem features, such as goals, objectives, opinions, etc. Decision makers then evaluate alternatives with respect to those problem features. This model is a key to combining the three parts of a decision basis. Utility—explicitly modelled or not—can be seen as an underlying basis of any rational choice. Often, the criteria for decision making are variables of the utility function, and the parameters indicate the importance of the criteria. A very simple utility model represents a decision consequence as the utility value U, which is the weighted (a_i's) sum of the decision criteria X_i evaluated on a particular alternative:

$$U = \sum_i a_i X_i. \tag{2.1}$$

The alternative that produces the highest utility value is accepted as having the most desirable outcome and, hence, should be the one selected. Typically, the approach is *normative* when the aim is neither to explain observed behaviour nor to predict how decisions will be made, but rather to facilitate better decisions than would be possible otherwise. Although human behaviour might not be explained using models of rational choice, preferences of decision makers can be analysed and decision alternatives can be evaluated based on those preferences by an analytical decision model (Kangas 1993). This process adds rigor to decision making and also makes it more explicit.

Utility is influenced by all attributes of the decision problem that have value to the decision maker. It is a measure of subjective desirability. Utility of a single decision maker can also include altruistic elements related to other people's preferences. In which case, maximizing one's expressed utility does not necessarily mean purely self-seeking behaviour. In most cases, utility cannot be expressed in physical quantities, e.g., monetary cost or benefit. The real utility of physical units is determined by their value to the decision maker, and it is, by no means, always linearly related to the units of physical quantities. In decision analyses, it is often better to use relative values instead of physical measures (Forman 1987). In the AHP, relative utility values are referred to as "priority," and the utility model as formulated in the AHP can be called a priority model.

If, by means of a priority model, decision alternatives can be arranged only from the best to the worst, one speaks of *ordinal* priority. If the priority model can be interpreted on an interval or a ratio scale, one speaks of *cardinal* utility. In principle, it is sufficient to determine the ordinal priorities only when the best decision alternative is sought. Estimating the cardinal utility, however, also enables a versatile analysis of a complex decision situation. Cardinality in a ratio scale, as applied in the AHP for instance, also enables sensitivity analysis and risk analyses, among other things, of the decision process (e.g., von Winterfeldt and Edwards 1988). This allows decision makers to conduct "what-if" scenarios and to evaluate the impact of uncertain preferences.

In most decision-making situations, the preferences of decision makers have been more or less neglected when alternatives are evaluated (e.g., Keeney 1988, Bradshaw and Boose 1990). This is also the case in natural resource management (Kangas 1992). For decision support based on operations research methods, problem structuring is too often technique oriented. When applying artificial intelligence methods, the decision-theoretic methodology is typically forgotten (O'Keefe 1988). This being the case, decision analyses in natural resource management can be improved significantly by developing and applying methods that place greater emphasis on the decision makers' and stakeholders' preferences when prioritising decision alternatives. The following section illustrates how the analytic hierarchy process offers an alternative approach to traditional operations research and normative decision methods.

2.2 The Analytic Hierarchy Process

Many decision-making situations involve preferential selection among alternative items, events, or courses of action. When the selection criterion is "least cost," the measurement scale is obvious and choosing becomes easy. In most real-world situations, however, there is not a single scale for measuring all competing alternatives. More often, there are several scales that must be used and often those scales are related to one another in fairly complex ways.

The AHP (Saaty 1980) is designed to help with multiple-criteria decisions. Three important components of the AHP are: (1) the structuring of a problem into a hierarchy consisting of a goal and subordinate features (decomposition), (2) pair-wise comparisons between elements at each level (evaluation), and (3) propagation of level-specific, local priorities to global priorities (synthesis). Subordinate levels of a hierarchy, may include: objectives, scenarios, events, actions, outcomes, and alternatives. Alternative courses of action to be compared appear at the lowest level of the hierarchy. Pair-wise comparisons are made between all elements at a particular level with respect to elements in the level above it. Comparisons can be made according to preference, importance, or likelihood—whichever is most appropriate for the elements considered. Saaty (1980) developed the mathematics to combine pairwise comparisons made at different levels in order to produce a final priority value for each of the alternatives at the bottom of the hierarchy.

As a simple and easily understood example, consider the hierarchy in Figure 1, which is designed to enable one to select a "best" college to attend. The goal, *satisfying college*, appears at the top of the hierarchy. The criteria appear on the next level: *academic reputation*, *cost*, *campus beauty*, *local living climate*, and *social life*. The colleges to be considered are labelled A, B, and C at the lowest level. First, the criteria are compared pair-wise with respect to their importance for producing a satisfying college experience. The scale of integers in the range 1-9 is used for comparison (Saaty 1990). One possible matrix resulting from these pairwise comparisons appears in Table 1. In this matrix, each value a_{ij} indicates how much more important, preferred, or likely row heading i is than column heading j. Corresponding matrix entries a_{ji} equal $1/a_{ij}$. Elements on the matrix diagonal are always unity. The normalized principal right eigenvector **c**' = [0.465, 0.326, 0.085, 0.097, 0.038] of this matrix represents the priority values of those criteria (Saaty 1980).

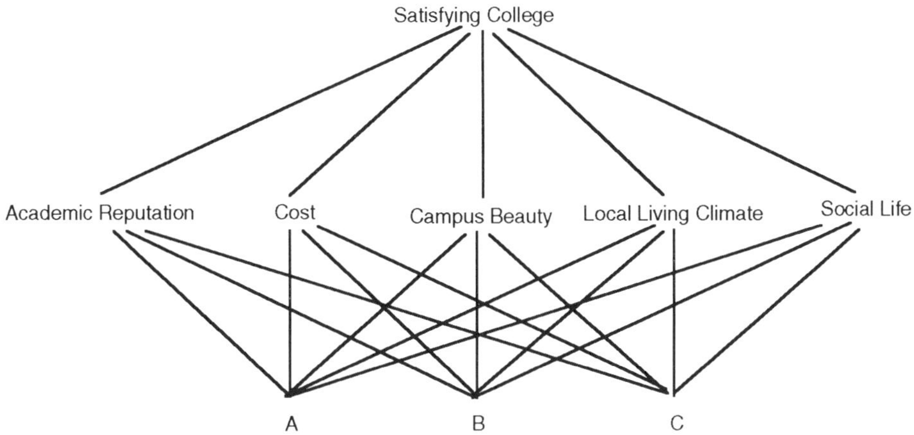

Figure 1. A simple analytic hierarchy for selecting a satisfying college from among three alternatives, A, B, and C, makes use of five criteria. Each of the alternative colleges is scored on each criteria. In general, however, a hierarchy need not be fully connected in this way.

Table 1. The five criteria for selecting a college are compared in a pairwise fashion and assigned a relative importance score.

	Academic Reputation	Cost	Campus Beauty	Local Living Climate	Social Life
Academic Reputation	1	3	5	3	7
Cost	1/3	1	5	5	9
Campus Beauty	1/5	1/5	1	1	3
Local Living Climate	1/3	1/5	1	1	3
Social Life	1/7	1/9	1/3	1/3	1

When all pair-wise comparisons in the judgment matrix **A** are absolutely consistent, i.e. $a_{ij}a_{jk}=a_{ik}$ for all $i\neq k$, then (2.2) holds, where **w** is the vector of priority values. This mathematical statement (2.2) also says that **w** is an eigenvector of **A** with associated eigenvalue n. Because the matrix multiplication occurs on the right, **w** is called a right eigenvector. In the consistent case, n is the only non-zero eigenvalue of $\mathbf{A}$. As judgments become inconsistent, however, small changes occur in the a_{ij}, and $\mathbf{A}$ becomes inconsistent. Then, multiple eigenvectors and eigenvalue solutions exist for (2.2). The largest (or principal) eigenvalue remains close to n as long as changes to the a_{ij} are small and $\mathbf{A}$ does not become too inconsistent (Saaty 1980). Therefore, the principal right eigenvector is still a good approximation to the consistent-case eigenvector $\mathbf{w}$.

$$\mathbf{Aw} = n\mathbf{w} \tag{2.2}$$

Then alternative colleges are compared regarding the extent to which each has these criteria. One matrix, such as Table 2, would be produced for each criterion. Similar to the first matrix (Table 1), a priority vector $\mathbf{w}_1' =$ [0.637, 0.258, 0.105] can be calculated from Table 2. Priority vectors $\mathbf{w}_2$, ..., $\mathbf{w}_5$ can also be generated for each of the remaining criteria. The degree to which the colleges possess each criterion (stored in the $\mathbf{w}_i$) is weighted by the importance of that criterion c_i and summed across all criteria to obtain a final priority value for that college. In matrix arithmetic, the final priority vector for the colleges is calculated as

$$\mathbf{w} = \left[\mathbf{w}_1\mathbf{w}_2\mathbf{w}_3\mathbf{w}_4\mathbf{w}_5\right]\mathbf{c} \tag{2.3}$$

A more detailed example of the AHP process appears in Schmoldt *et al.* (1994) with some of the mathematical derivations. Because the final result of the AHP is a numerical priority value for each alternative, the decision maker may then select the highest scoring alternative as the "best." The decision process that has been made explicit in the hierarchy and in the comparisons determines this "best" alternative.

Table 2. The three colleges are compared with respect to the criterion, academic reputation.

Academic Reputation	College A	College B	College C
College A	1	3	5
College B	1/3	1	3
College C	1/5	1/3	1

The analytic hierarchy process has been applied to a wide variety of decision-making problems, both in a practical, as well as academic, context (Zahedi 1986). For example, it has been used for planning, resource allocation, and priority setting in business, energy, health, marketing, forest management, and transportation. The AHP is relevant to nearly any natural resource/environmental management application that requires multiple opinions, multiple participants, or a complex, decision-making process. The next section highlights a few of the many such AHP applications.

3. THE AHP AND NATURAL RESOURCE MANAGEMENT

Because natural resource management often entails making choices among alternative management regimes, decision support tools are proposed as instruments for making rational, carefully reasoned, and justifiable decisions. This section briefly reviews some of the applications of these decision support tools, particularly the AHP, for forestry and natural resources. The review does not focus on technical issues; the chapters contained in this book offer excellent expositions on both the technical aspects of the method and novel approaches used to apply the method to different problem situations.

While the AHP was developed only in the late 1970's, it has become one of the most widely used techniques as shown by the extensive literature published in journals and books, most of which are in areas outside natural resources. AHP applications in forestry, agriculture, and natural resources are still surprisingly limited. Chapter contributions contained in this book constitute perhaps the most updated compendium of recent applications of the AHP in natural resource and environmental management. These chapters also contain extensive reviews of literature that may not be covered in this section.

Published applications in forestry include: forest management (Mendoza and Sprouse 1989); forest planning and decision making (Kangas *et al.* 1996, Pukkala and Kangas 1993); risk assessment in assessing reforestation alternatives (Kangas 1993a); risk and attitude toward risks in forest planning (Pukkala and Kangas 1996); eco-labelling and certification of forest products (Pesonen *et al.* 1997); forest protection through selection of risk factors for spruce beetle outbreaks (Reynolds and Holsten 1994); setting priorities for restoration projects (Reynolds 1997); identification and prioritisation of fire research needs (Schmoldt and Peterson 2000); and assessment of criteria and indicators for evaluating forest sustainability (Mendoza and Prabhu 2000). Other forest-related applications of the AHP include: assessment of forests' scenic values (Kangas *et al.* 1992); assessment of factors affecting timber bridge materials (Smith *et al.* 1995); development of resource management plans for National Parks (Peterson *et al.* 1994); and resource inventory and monitoring in National Parks (Schmoldt *et al.* 1994).

Wildlife management is another area that has received considerable attention for AHP-related studies. Pereira and Duckstein (1993) combined the AHP with geographic information system (GIS) to study habitat suitability for Mount Graham red squirrel. Mendoza (1997) also described an integrated model combining the AHP with GIS to generate habitat suitability indices for desert tortoise. Kangas *et al.* (1993b) used the AHP to estimate wildlife habitat suitability functions using experts' judgments.

Other applications include: measurement of consumer preferences for environmental policy (Uusitalo 1990); evaluation of irrigation systems (Mingyao 1994); managing fisheries (DiNardo *et al.* 1989, Imber 1989, Levy 1989); energy planning and resource allocation (Hamalainen and Seppalainen 1986, Gholamnezhad and Saaty 1982); and sustainable agriculture (Mawampanga 1993).

One of the areas where the AHP has received wide application is land use suitability analysis. Banai-Kashani (1989) and Xiang and Whitley (1994) offer excellent reviews describing the potential of the AHP for general site suitability and land capability analyses. Huchinson and Toledano (1993) describe the use of the AHP in conjunction with GIS for designing land use plans considering multiple objectives and participatory approaches to planning and decision making. As land use become more constrained and the land allocated to various activities continues to shrink, suitability analyses take on added importance.

4. CONCLUSIONS

The days are long gone when natural resource decisions could be based on a single metric, e.g. net present monetary value, while addressing a single resource, e.g., timber. Even the decision-making protocol has changed, now including multiple participants with vastly different value systems. Normative decision methods (offering a rational choice) must now include both decision makers and stakeholders, and must quantify their preferences in a realistic way.

The analytic hierarchy process not only offers some advantages over traditional decision methods, but it can integrate with those other approaches to take advantage of the strengths inherent in each. Several AHP applications are mentioned above, while the remainder of this text provides many detailed examples. Even though the number of AHP applications described in forestry and related disciplines is growing steadily, real-world examples of the AHP in actual resource management use are extremely limited. Given the method's relative ease of use, and yet broad applicability, its disuse is somewhat surprising. In our experience, though, it seems that many land management organizations expend a great deal of time and effort collecting information about managed resources, decision alternatives, and decision consequences, but pay relatively little attention to how all that information must be integrated into a rational choice. The assumption seems to be that the correct decision alternative will materialize automatically from enormous data gathering efforts. Rather, a decision framework, like multi-attribute utility theory or the AHP, is the glue that binds all of the decision support information together, and helps the decision maker create some sense out of it. Even with volumes of information, there is no guarantee that good decisions will result. Significant effort must also be placed on how preferential choices are made.

Considering the complexity of most management issues and compliance regulations, the AHP can extend to a wide array of managerial and planning tasks. For example, management and planning for a large watershed may include issues related to water quality and quantity, forest management, wildlife management, and recreation. Input is required from subject matter experts in each of these disciplines in order to establish priorities and make informed decisions regarding spatial and temporal distributions of resources. Because watersheds generally involve the flow of materials between public and private lands, additional input is often needed on social, legal, and political aspects of resource condition and value. In addition to its breadth of application, the AHP is relatively easy to apply, to understand, and to interpret. These attributes of the AHP validate its focus in this book as a valuable tool for decision making.

REFERENCES

BANAI-KASHANI, R. 1989. A new method for site suitability analysis: The analytic hierarchy process. Environmental Management 13: 685-693.

BRADSHAW, J. M. AND BOOSE, J. H. 1990. Decision analysis techniques for knowledge acquisition: combining information and preferences using Aquinas and Axotl. International Journal of Man-Machine Studies 32: 121-186.

DINARDO, G., LEVY, D. AND GOLDEN, B. L. 1989. Using decision analysis to manage Maryland's River Herring Fishery: An application of the AHP. Journal of Environmental Management 29: 193-213.

FEMAT. 1993. Forest Ecosystem Management: An Ecological, Economic, and Social Assessment. U.S. Department of Agriculture, Forest Service, Washington D.C.

FORMAN, E. H. 1987. Relative vs. absolute worth. Mathematical Modelling 9: 195-202.

GHOLAMNEZHAD, A. AND SAATY, T. L. 1982. A desired energy mix for the United States in the year 2000: An analytic hierarchy process approach. Int. J. of Policy Analysis and Information Systems 6(1): 47-64.

GUARISO, G. AND WERTHNER, H. 1989. Environmental decision support systems. Ellis Horwood, Chichester.

HAMALAINEN, R. P. AND SEPPALAINEN, T. O. 1986. The analytic hierarchy process in energy policy planning. Socio-Economic Planning Sci. 20: 399-405.

HUCTHINSON, C. F. AND TOLEDANO, J. 1993. Guidelines for demonstrating geographical information systems based on participatory development. Int. J. Geographical Information Systems 7: 453-461.

IMBER, S. E. 1989. Applications of linear programming and the analytic hierarchy process to fisheries management. M.S. Thesis. University of Maryland.

KANGAS, J. 1992a. Metsikön uudistamisketjun valinta - monitavoitteiseen hyötyteoriaan perustuva päätösanalyysimalli. Summary: Choosing the regeneration chain in a forest stand: A decision model based on multi-attribute utility theory. Doctoral thesis. University of Joensuu, Publications in Sciences 24.

KANGAS, J. 1992b. Multiple-use planning of forest resources by using the analytic hierarchy process. Scand. J. of For. Research 7: 259-268.

KANGAS, J. 1993. A multi-attribute preference model for evaluating the reforestation chain alternatives of a forest stand. Forest Ecology and Management 59: 271-288.

KANGAS, J., KARSIKKO, J., LAASONEN, L. AND PUKKALA, T. 1993. A method for estimating the suitability function of wildlife habitat for forest planning on the basis of expertise. Silva Fennica 27: 259-268.

KANGAS, J., LOIKKANEN, T., PUKKALA, T. AND PYKALAINEN, J. 1992. A participatory approach to tactical forest planning. Acta Forestalia Fennica 215.

KEENEY, R. L. 1982. Decision analysis: An overview. Operations Research 30: 803-838.

KEENEY, R. L. 1988. Value-driven expert systems for decision support. Decision Support Systems 4: 405-412.

LACKEY, R. T. 1998. Seven pillars of ecosystem management. Landscape and Urban Planning 40: 21-30.

LEVY, D. 1989. The analytic hierarchy process: applications to natural resources and environmental management. M.S. Thesis. University of Maryland.

MAWAMPANGA, M. 1993. An application of the analytic hierarchy process to the farmer's decision to adopt sustainable agriculture. Ph.D. Thesis. University of Kentucky.

MENDOZA, G. A. 1997. Introduction to analytic hierarchy process: Theory and applications to natural resources management. Pages 130-139 in 1997 ACSM/ASPRS/RT Convention, Vol. 4 Resource Technology. American Society of Photogrammetry and Remote Sensing, Bethesda MD.

MENDOZA, G. A. AND PRABHU, R. 2000. Multiple criteria decision-making approaches to assessing forest sustainability using criteria and indicators: A case study. Forest Ecology and Management 131: 107-126.

MENDOZA, G. A. AND SPROUSE, W. 1989. Forest planning and decision making under fuzzy environments: An overview and analysis. Forest Science 35: 481-502.

MINGYAO, Z. 1994. Group analytic hierarchy process (GAHP) -- fuzzy method for evaluation of irrigation district management. Irrigation and Drainage Systems 8: 177-188.

O'KEEFE, R. M. 1988. Artificial intelligence and the management science practitioner: Expert systems and MS/OR methodology (Good News and Bad). Interfaces 18: 105-113.

PEREIRA, J. M. AND DUCKSTEIN, L. 1991. GIS-based habitat modeling using logistic multiple regression: A study of the Mt. Graham Red Squirrel. Photogrammetric Engineering and Remote Sensing 57: 1475-1486.

PESONEN, M., KAJANUS, M. AND KURTTILA, M. Using AHP in analysing results of SWOT: A case of eco-labeled forestry in Finland. Pages 171-179 in 1997 ACSM/ASPRS/RT Convention, Vol. 4 Resource Technology. American Society of Photogrammetry and Remote Sensing, Bethesda MD.

PETERSON, D. L., SILSBEE, D. AND SCHMOLDT, D. L. 1994. A case study of resource management planning with multiple objectives. Environmental Management 18: 729-742.

PINCHOT, G. 1947. Breaking New Ground. Univ. of Wash. Press, Seattle.

PUKKALA, T. AND KANGAS, J. 1996. A method of integrating risk and attitude toward risk into forest planning. Forest Science 42: 198-205.

REYNOLDS, K. 1997. Setting priorities for maintenance and restoration projects with the analytic hierarchy process and SMART. Pages 163-170 in 1997 ACSM/ASPRS/RT Convention, Vol. 4 Resource Technology. American Society of Photogrammetry and Remote Sensing, Bethesda MD.

REYNOLDS, K., BJORK, J., RIEMANN-HERSHEY, R., SCHMOLDT, D., PAYNE, J., KING, S., MOEUR, M., DECOLA, L., TWERY, M. AND CUNNINGHAM, P. 1999. Chapter 28: Decision support for ecosystem management. Pages 687-721 in Johnson, N. C., Malk, A. J., Sexton, W. T. and Szaro, R. (eds.) Ecological Stewardship: A Common Reference for Ecosystem Management. Elsevier Science Ltd., Oxford.

REYNOLDS, K. M. AND HOLSTEN, E. 1994. Relative importance of risk factors for spruce beetle outbreaks. Can. J. of For. Research. 24: 2089-2095.

SAATY, T. L. 1980. The Analytic Hierarchy Process. McGraw Hill, New York.

SAATY, T. L. 1990. Multicriteria Decision Making: The Analytic Hierarchy Process. RWS Publications, Pittsburgh.

SCHMOLDT, D.L. AND PETERSON, D.L. 2000. Analytical group decision making in natural resources: Methodology and application. Forest Science 46: 62-75.

SCHMOLDT, D. L., PETERSON, D. L. AND SILSBEE, D. 1994. Developing inventory and monitoring programs based on multiple objectives. Environmental Management 18(5): 707-727.

SCHMOLDT, D. L. AND RAUSCHER, H. M. 1996. Building Knowledge-Based Systems for Natural Resource Management. Chapman & Hall, New York.

SMITH, R. L., BUSH, R., SCHMOLDT, D. L. 1995. A hierarchical model and analysis of factors affecting the adoption of timber as a bridge material. Wood and Fiber Science 27: 225-238.

THOMAS, D.A. AND SUN, X. 1995. Rangeland production: Use of models incorporating aggregated knowledge and fuzzy construction. Journal of Arid Environments 30: 479-494.

UNGER, D. AND SALWASSER, H. 1991. Balancing people's needs and resource quality on the national forests and grasslands. Pages 47-51 in Agriculture and the Environment. The 1991 Yearbook of Agriculture, U.S. Department of Agriculture, Washington DC.

UUSITALO, L. 1990. Consumer preferences for environmental quality and other social goals. Journal of Consumer Policy 13: 231-251.

VON WINTERFELDT, D. 1988. Expert systems and behavioral decision research. Decision Support Systems 4: 461-471.

VON WINTERFELDT, D. AND EDWARDS, W. 1988. Decision analysis and behavioral research. Cambridge University Press, Cambridge.

XIANG, W. AND WHITLEY, D. L. 1994. Weighting land suitability factors by the PLUS Method. Environment and Planning B: Planning and Design 21: 273-304.

ZAHEDI, F. 1986. The analytic hierarchy process—A survey of the method and its applications. Interfaces 16: 96-108.

Chapter 2

Fundamentals of the Analytic Hierarchy Process

Thomas L. Saaty
University of Pittsburgh, Pittsburgh PA USA

Key words: Analytic hierarchy process, ratio scale, subjective judgement, group decision making.

Abstract: The seven pillars of the analytic hierarchy process (AHP) are presented. These include: (1) ratio scales derived from reciprocal paired comparisons; (2) paired comparisons and the psychophysical origin of the fundamental scale used to make the comparisons; (3) conditions for sensitivity of the eigenvector to changes in judgements; (4) homogeneity and clustering to extend the scale from 1-9 to 1-∞; (5) additive synthesis of priorities, leading to a vector of multi-linear forms as applied within the decision structure of a hierarchy or the more general feedback network to reduce multi-dimensional measurements to a uni-dimensional ratio scale; (6) allowing rank preservation (ideal mode) or allowing rank reversal (distributive mode); and (7) group decision making using a mathematically justifiable way for synthesising individual judgements which allows the construction of a cardinal group decision compatible with individual preferences. These properties of the AHP give it both theoretical support and broad application.

1. INTRODUCTION

The analytic hierarchy process (AHP) provides the objective mathematics to process the inescapably subjective and personal preferences of an individual or a group in making a decision. With the AHP and its generalisation, the analytic network process (ANP), one constructs hierarchies or feedback networks that describe the decision environment structure. The decision maker then makes judgements or performs measurements on pairs of elements with respect to a controlling element to derive ratio scales that are then synthesised throughout the structure to select the best alternative.

D.L. Schmoldt et al. (eds.),
The Analytic Hierarchy Process in Natural Resource and Environmental Decision Making, 15–35.

Fundamentally, the AHP works by developing priorities for alternatives and the criteria used to judge the alternatives. Criteria are selected by a decision maker (irrelevant criteria are those that are not included in the hierarchy). Selected criteria may be measured on different scales, such as weight and length, or may even be intangible for which no scales yet exist. Measurements on different scales, of course, cannot be directly combined. First, priorities are derived for the criteria in terms of their importance to achieve the goal, then priorities are derived for the performance of the alternatives on each criterion. These priorities are derived based on pairwise assessments using judgement or ratios of measurements from a scale if one exists. The process of prioritisation solves the problem of having to deal with different types of scales, by interpreting their significance to the values of the user or users. Finally, a weighting and adding process is used to obtain overall priorities for the alternatives as to how they contribute to the goal. This weighting and adding parallels what one would have done arithmetically prior to the AHP to combine alternatives measured under several criteria having the *same* scale to obtain an overall result (a scale that is often common to several criteria is money). With the AHP a multidimensional scaling problem is thus transformed to a uni-dimensional scaling problem.

The AHP can be viewed as a formal method for rational and explicit decision making. It possesses the seven fundamental properties, below. Subsequent sections examine each in greater detail.

Normalised ratio scales are central to the generation and synthesis of priorities, whether in the AHP or in any multicriteria method that needs to integrate existing ratio scale measurements with its own derived scales.

Reciprocal paired comparisons are used to express judgements semantically, and to automatically link them to a numerical and fundamental scale of absolute numbers (derived from stimulus-response relations). The principal right eigenvector of priorities is then derived; the eigenvector shows the dominance of each element with respect to the other elements. Inconsistency in judgement is allowed and a measure for it is provided which can direct the decision maker in both improving judgement and arriving at a better understanding of the problem. The AHP has at least three modes for arriving at a ranking of the alternatives: *relative*, which *ranks* a few alternatives by comparing them in pairs (particularly useful in new and exploratory decisions), *absolute*, which *rates* an unlimited number of alternatives one at a time on intensity scales constructed separately for each covering criterion (particularly useful in decisions where there is considerable knowledge to judge the relative importance of the intensities), and *benchmarking*, which ranks alternatives by including a known alternative in the group and comparing the others against it.

Sensitivity of the principal right eigenvector to perturbation in judgements limits the number of elements in each set of comparisons to a few and requires that they be homogeneous.

Homogeneity and clustering are used to extend the fundamental scale gradually from cluster to adjacent cluster, eventually enlarging the scale from 1-9 to 1-∞.

Synthesis that can be extended to dependence and feedback is applied to the derived ratio scales to create a uni-dimensional ratio scale for representing the overall outcome. Synthesis of the scales derived in the decision structure can only be made to yield correct outcomes on known scales by additive weighting.

Rank preservation and reversal can be shown to occur without adding or deleting criteria, such as by simply introducing enough copies of an alternative. This leaves no doubt that rank reversal is as intrinsic to decision making as rank preservation also is.

Group judgements must be integrated one at a time carefully and mathematically, taking into consideration, when desired, the experience, knowledge, and power of each person involved in the decision. The AHP's *cardinal* ratio scale preferences allow one the *possibility* of constructing a social utility function—an impossibility when using *ordinal* preferences. To deal with a large group requires the use of questionnaires and statistical procedures for large samples.

2. RATIO SCALES

A *ratio* is the relative value or quotient a/b of two quantities a and b *of the same kind*; it is called commensurate if it is a rational number, otherwise it is incommensurate. A statement of the equality of two ratios a/b and c/d is called *proportionality*. A ratio scale is a set of numbers that is invariant under a similarity transformation (multiplication by a positive constant). The constant cancels when the ratio of any two numbers is formed. Either pounds or kilograms can be used to measure weight, but the ratio of the weight of two objects is the same for both scales. An extension of this idea is that the weights of an entire set of objects, whether in pounds or in kilograms, can be standardised to read the same by normalising. In general if the readings from a ratio scale are aw_i^*, $i=1,\ldots,n$, the standard form is given by $w_i=aw_i^*/aw_i^*=w_i^*/w_i^*$ as a result of which we have $\Sigma w_i=1$, and the w_i, $i=1,\ldots,n$, are said to be normalised. We no longer need to specify whether weight for example is given in pounds or in kilograms or in another kind of unit. The weights (2.21, 4.42) in pounds and (1, 2) in kilograms, are both given by (1/3, 2/3) in the standard ratio scale form.

The relative ratio scale derived from a pairwise comparison reciprocal matrix of judgements is derived by solving:

$$\begin{cases} \sum_{j=1}^{n} a_{ij} w_j = \lambda_{max} w_i \\ \sum_{i=1}^{n} w_i = 1 \end{cases} \tag{2.1}$$

with $a_{ji}=1/a_{ij}$ or $a_{ij}a_{ji}=1$ (the reciprocal property), $a_{ij}>0$ (thus $\mathbf{A}$ is known as a positive matrix) whose solution, known as the principal right eigenvector, is normalised. A relative ratio scale does not need a unit of measurement.

When $a_{ij}a_{jk} = a_{ik}$, the matrix $\mathbf{A}=(a_{ij})$ is said to be consistent and its principal eigenvalue is equal to n. Otherwise, it is simply reciprocal. The general eigenvalue formulation given in (2.1) is obtained by perturbation of the following consistent formulation:

$$\mathbf{Aw} = \begin{array}{c} \\ \mathbf{A}_1 \\ \vdots \\ \mathbf{A}_n \end{array} \begin{array}{c} \begin{array}{ccc} \mathbf{A}_1 & \cdots & \mathbf{A}_n \end{array} \\ \begin{bmatrix} \frac{w_1}{w_1} & \cdots & \frac{w_1}{w_n} \\ \vdots & \ddots & \vdots \\ \frac{w_n}{w_1} & \cdots & \frac{w_n}{w_n} \end{bmatrix} \end{array} \begin{bmatrix} w_1 \\ \vdots \\ w_n \end{bmatrix} = n \begin{bmatrix} w_1 \\ \vdots \\ w_n \end{bmatrix} = n\mathbf{w} \tag{2.2}$$

where $\mathbf{A}$ has been multiplied on the right by the transpose of the vector of weights $\mathbf{w}=(w_1,\ldots,w_n)$. The result of this multiplication is $n\mathbf{w}$. Thus, to recover the scale from the matrix of ratios, one must solve the problem $\mathbf{Aw}=n\mathbf{w}$ or $(\mathbf{A}-n\mathbf{I})\mathbf{w} = \mathbf{0}$. This is a system of homogeneous linear equations. It has a nontrivial solution if and only if the determinant of $\mathbf{A}-n\mathbf{I}$ vanishes, that is, n is an eigenvalue of $\mathbf{A}$. Now $\mathbf{A}$ has unit rank since every row is a constant multiple of the first row. Thus, all its eigenvalues except one are zero. The sum of the eigenvalues of a matrix is equal to its trace, that is, the sum of its diagonal elements. In this case, the trace of $\mathbf{A}$ is equal to n. Thus n is an eigenvalue of $\mathbf{A}$, and one has a nontrivial solution. The solution consists of positive entries and is unique to within a multiplicative constant.

The discrete formulation given in (2.1) above generalises to the continuous case through Fredholm's integral equation of the second kind and is given by:

$$\int_a^b K(s,t)w(t)dt = \lambda_{max} w(s), \quad \int_a^b w(s)ds = 1 \tag{2.3}$$

where instead of the matrix **A** we have as a positive kernel, $K(s,t) > 0$. Note that the entries in a matrix **A** depend on the two variables i and j which assume discrete values. Thus, the matrix itself depends on these discrete variables, and its generalisation, the kernel function, also depends on two (continuous) variables. The reason for calling it a kernel is the role it plays in the integral, where we cannot determine the exact form of the solution without knowing the kernel. The standard reciprocal form of (2.3) is written by moving the eigenvalue to the left hand side. As in the finite case, we have the reciprocal property and the consistency relation (2.4).

$$\begin{cases} K(s,t)K(t,s) = 1 \\ K(s,t)K(t,u) = K(s,u) \quad \forall s,t,u \end{cases} \tag{2.4}$$

An example of this type of kernel is $K(s,t)=e^{s-t}=e^s/e^t$. It follows by putting $s=t=u$, that $K(s,s)=1$ for all s which is analogous to having ones down the diagonal of the matrix in the discrete case. A value of λ for which Fredholm's equation has a nonzero solution $w(t)$ is called a characteristic value (or its reciprocal is called an eigenvalue) and the corresponding solution is called an eigenfunction. An eigenfunction is determined to within a multiplicative constant. If $w(t)$ is an eigenfunction corresponding to the characteristic value λ and if C is an arbitrary constant, we can easily see by substituting in the equation that $Cw(t)$ is also an eigenfunction corresponding to the same λ. The value $\lambda=0$ is not a characteristic value because we have the corresponding solution $w(t)=0$ for every value of t, which is the trivial case, excluded in our discussion.

A matrix is consistent if and only if it has the form $\mathbf{A}=(w_i/w_j)$ which is equivalent to multiplying a column vector that is the transpose of $(w_1, \ldots, w_n)$ by the row vector $(1/w_1, \ldots, 1/w_n)$. As we see below, the kernel $K(s,t)$ is separable and can be written as

$$K(s,t) = k_1(s)k_2(\mathrm{t}) \tag{2.5}$$

Theorem *K(s,t) is consistent if and only if it is separable of the form:*

$$K(s,t) = k(s)\,/\,k(\mathrm{t}) \tag{2.6}$$

Theorem *If K(s,t) is consistent, the solution of (2.3) is given by*

$$w(s) = \frac{k(s)}{\int_S k(s)ds} \tag{2.7}$$

In the discrete case, the normalised eigenvector is independent of whether all the elements of the pairwise comparison matrix **A** are multiplied by the same constant a or not, and thus we can replace **A** by a**A** and obtain the same eigenvector. Generalising this result we have:

$$K(as,at) = aK(s,t) = k(as) / k(at) = ak(s) / k(t) \tag{2.8}$$

which means that K is a homogeneous function of order one. In general, when $f(ax_1, \ldots, ax_n)=a^n f(x_1, \ldots, x_n)$ holds, f is said to be homogeneous of order n. Because K is a degenerate kernel, we can replace $k(s)$ above by $k(as)$ and obtain $w(as)$. We have now derived from considerations of ratio scales the following condition to be satisfied by a ratio scale:

Theorem *A necessary and sufficient condition for w(s) to be an eigenfunction solution of Fredholm's equation of the second kind, with a consistent kernel that is homogeneous of order one, is that it satisfy the functional equation* .

$$w(as) = bw(s), \text{ where } b = \alpha a. \tag{2.9}$$

We have for the general damped periodic response function $w(s)$,

$$w(s) = Ce^{\log b\left(\frac{\log s}{\log a}\right)} P\left(\frac{\log s}{\log a}\right) \tag{2.10}$$

where P is periodic of period 1 and $P(0)=1$.

We can write this solution as

$$v(u) = C_1 e^{-\beta u} P(u) \tag{2.11}$$

where $P(u)$ is periodic of period 1, $u=\log s/\log a$ and $\log ab=-\beta$, $\beta>0$. It is interesting to observe the logarithmic function appear as part of the solution. It gives greater confirmation to the Weber-Fechner law developed in the next section.

3. PAIRED COMPARISONS AND THE FUNDAMENTAL SCALE

Instead of assigning two numbers w_i and w_j and forming the ratio w_i/w_j we assign a single number drawn from the fundamental 1-9 scale of absolute numbers to represent the ratio $(w_i/w_j)/1$. It is a nearest integer approximation to the ratio w_i/w_j. The derived scale will reveal what the w_i and w_j are. This is a central fact about the relative measurement approach of the AHP and the need for a fundamental scale.

In 1846, Weber found, for example, that people holding different weights in their hand, could distinguish between a weight of 20 g and a weight of 21 g, but could not if the second weight is only 20.5 g. On the other hand, while they could not distinguish between 40 g and 41 g, they could between 40 g and 42 g, and so on at higher levels. We need to increase a stimulus s by a minimum amount Δs to reach a point where our senses can first discriminate between s and $s+\Delta s$. The amount Δs is called the just noticeable difference (jnd). The ratio $r=\Delta s/s$ does not depend on s. Weber's law states that change in sensation is noticed when the stimulus is increased by a constant percentage of the stimulus itself. This law holds in ranges where Δs is small when compared with s, and hence in practice it fails to hold when s is either too small or too large. Aggregating or decomposing stimuli as needed into clusters or hierarchy levels is an effective way to extend the use of this law.

In 1860, Fechner considered a sequence of just noticeable increasing stimuli. He denotes the first one by s_0. The next just noticeable stimulus is given by

$$s_1 = s_0 + \Delta s_0 = s_0 + \frac{\Delta s_0}{s_0} s_0 = s_0(1+r) \tag{3.1}$$

based on Weber's law. Similarly,

$$s_2 = s_1 + \Delta s_1 = s_1(1+r) = s_0(1+r)^2 \equiv s_0\alpha^2. \tag{3.2}$$

In general,

$$s_n = s_{n-1}\alpha = s_0\,\alpha^n \quad (n = 0,\ 1,\ 2,\ \ldots). \tag{3.3}$$

Thus, stimuli of noticeable differences follow sequentially in a geometric progression. Fechner noted that the corresponding sensations should follow each other in an arithmetic sequence at the discrete points at which just noticeable differences occur. However, the latter are obtained when we solve for n. We have

$$n = \frac{(\log s_n \text{ - } \log s_0)}{\log \alpha} \tag{3.4}$$

and sensation is a linear function of the logarithm of the stimulus. Thus, if M denotes the sensation and s the stimulus, the psychophysical law of Weber-Fechner is given by

$$M = a \log s + b, \ a \neq 0. \tag{3.5}$$

We assume that the stimuli arise in making pairwise comparisons of relatively comparable activities. We are interested in responses whose numerical values are in the form of ratios. Thus b=0, from which we must have log s_0=0 or s_0=1, which is possible by calibrating a unit stimulus. Here the unit stimulus is s_0. The next noticeable stimulus is $s_1 = s_0\alpha = \alpha$ which yields the second noticeable response $a(\log\alpha)$. The third noticeable stimulus is s_2=$s_0\alpha^2$ which yields a response of $2a(\log\alpha)$. Thus, we have for the different responses:

$$M_0 = a \log s_0, \ M_1 = a \log \alpha, \ M_2 = 2a \log \alpha, \ \ldots, \ M_n = na \log \alpha. \tag{3.6}$$

While the noticeable ratio stimulus increases geometrically, the response to that stimulus increases arithmetically. Note that M_0=0 and there is no response. By dividing each M_i by M_1 we obtain the sequence of absolute numbers 1, 2, 3, ... of the fundamental 1-9 scale. Paired comparisons are made by identifying the less dominant of two elements and using it as the unit of measurement. One then determines, using the scale 1-9 or its verbal equivalent, how many times more the dominant member of the pair is than this unit. In making paired comparisons, we use the nearest integer approximation from the scale, relying on the insensitivity of the eigenvector to small perturbations (discussed below). The reciprocal value is then automatically used for the comparison of the less dominant element with the more dominant one. Despite the foregoing derivation of the scale in the form of integers, someone might think that other scale values would be better, for example using 1.3 in the place of 2. Imagine comparing the magnitude of two people with respect to the magnitude of one person and using 1.3 for how many instead of 2.

We note that there may be elements that are closer than 2 on the 1-9 scale, and we need a variant of the foregoing. Among the elements that are close, we select the smallest. Observe the incremental increases between that smallest one and the rest of the elements in the close group. We now consider these increments to be new elements and pairwise compare them on the scale 1-9. If two of the increments are themselves closer than 2 we treat them as identical, assigning a 1 (we could carry this on ad infinitum). In the end, each component of the eigenvector of comparisons for the increments is added to unity to yield the un-normalised priorities of the close elements for that criterion. Note that only the least of these close elements is used in comparisons with the other elements that can be compared directly using the normal 1-9 scale. Its priority is used to multiply the priorities of these close elements and finally the priorities of all the elements are re-normalised.

How large should the upper value of the scale be? Qualitatively, people have a capacity to divide their response to stimuli into three categories: high, medium and low. They also have the capacity to refine this division by further subdividing each of these intensities of responses into high, medium and low, thus yielding in all nine subdivisions. It turns out, from the requirement of homogeneity developed below, that to maintain stability (and limit inconsistency), our minds work with a few elements at a time.

4. SENSITIVITY OF THE PRINCIPAL EIGENVECTOR

To a first order approximation, perturbation $\Delta\mathbf{w}_1$ in the principal eigenvector $\mathbf{w}_1$ from perturbation $\Delta\mathbf{A}$ in the consistent matrix $\mathbf{A}$ is given by:

$$\Delta\mathbf{w}_1 = \sum_{j=2}^{n} (\mathbf{v}_j^{\mathrm{T}} \Delta\mathbf{A}\mathbf{w}_1 / (\lambda_1 - \lambda_j)\mathbf{v}_j^{\mathrm{T}}\, \mathbf{w}_j)\, \mathbf{w}_j \tag{4.1}$$

The eigenvector $\mathbf{w}_1$ is insensitive to perturbation in $\mathbf{A}$, if the principal eigenvalue λ_1 is separated from the other eigenvalues λ_j, here assumed to be distinct, and none of the products $\mathbf{v}_j^{\mathrm{T}}\mathbf{w}_j$ of left and right eigenvectors is small. We should recall that the nonprincipal eigenvectors need not be positive in all components, and they may be complex. One can show that all the $\mathbf{v}_j^{\mathrm{T}}\mathbf{w}_j$ are of the same order, and that $\mathbf{v}_j^{\mathrm{T}}\mathbf{w}_j$, the product of the normalised left and right principal eigenvectors, is equal to n. If n is relatively small and the elements being compared are homogeneous, none of the components of $\mathbf{w}_1$ is arbitrarily small and correspondingly, none of the components of $\mathbf{v}_1^{\mathrm{T}}$ is arbitrarily small. Their product cannot be arbitrarily small, and thus $\mathbf{w}$ is insensitive to small perturbations of the consistent matrix $\mathbf{A}$. The conclusion is that n must be small, and one must compare homogeneous elements.

5. CLUSTERING TO EXTEND THE SCALE FROM 1-9 TO 1-∞

In Figure 1, an unripe cherry tomato is eventually and indirectly compared with a large watermelon by first comparing it with a small tomato and a lime, the lime is then used again in a second cluster with a grapefruit and a honey dew where we then divide by the weight of the lime and then multiply by its weight in the first cluster, and then use the honey dew again in a third cluster and so on. In the end we have a comparison of the unripe cherry tomato with the large watermelon and would accordingly extended the scale from 1-9 to 1-721.

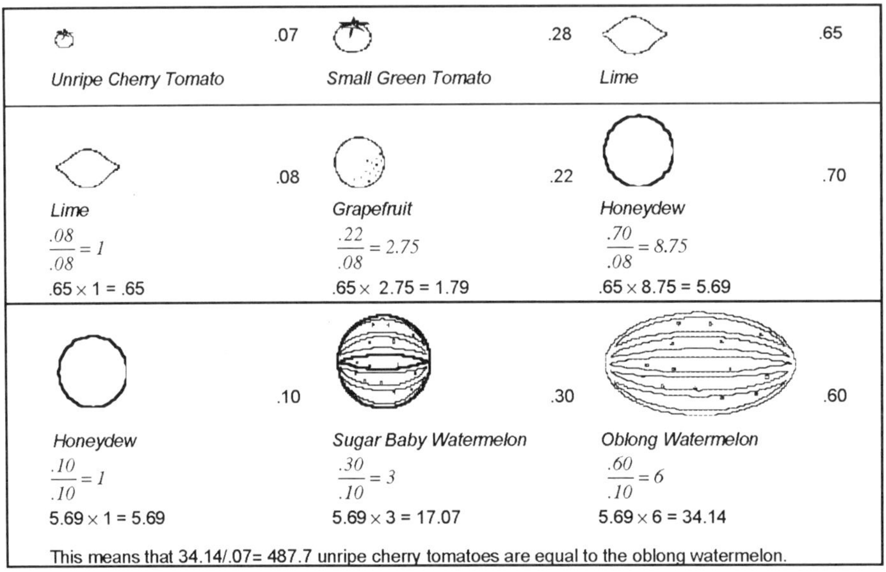

Figure 1. Comparisons according to volume.

Such clustering is essential, and must be done separately for each criterion. We should note that in most decision problems, there may be one or two levels of clusters and conceivably it may go up to three or four adjacent ranges of homogeneous elements (Maslow put them in seven groupings). Very roughly we have in decreasing order of importance: (1) survival, health, family, friends and basic religious beliefs some people were known to die for; (2) career, education, productivity and lifestyle; (3) political and social beliefs and contributions; (4) beliefs, ideas, and things that are flexible and it does not matter exactly how one advocates or uses them. These categories can be generalised to a group, a corporation, or a government. For very important decisions, two categories may need to be considered. Note that the priorities in two adjacent categories would be sufficiently different, one being an order of magnitude smaller than the other, that in the synthesis, the priorities of the elements in the smaller set have little effect on the decision. We do not have space to show how some *undesirable* elements can be compared among themselves and gradually extended to compare them with *desirable* ones as above. Thus one can go from negatives to positives but keep the measurement of the two types positive, by eventually clustering them separately.

6. SYNTHESIS: HOW TO COMBINE TANGIBLES WITH INTANGIBLES – ADDITIVE VS MULTIPLICATIVE

Let H be a complete hierarchy with h levels. Let $\mathbf{B}_k$ be the priority matrix of the kth level, $k=2,\ldots,h$. If $\mathbf{W}'$ is the global priority vector of the pth level with respect to some element z in the $(p-1)$st level, then the priority vector $\mathbf{W}$ of the qth level $(p<q)$ with respect to z is given by the multilinear (and thus nonlinear) form,

$$\mathbf{W} = \mathbf{B}_q \mathbf{B}_{q-1} \cdots \mathbf{B}_{p+1} \mathbf{W}'. \tag{6.1}$$

The global priority vector of the lowest level with respect to the goal is given by,

$$\mathbf{W} = \mathbf{B}_h \mathbf{B}_{h-1} \cdots \mathbf{B}_2 \mathbf{W}'. \tag{6.2}$$

In general, $\mathbf{W}'$ equals 1. The sensitivity of the bottom level alternatives with respect to changes in the weights of elements in any level can be studied by means of this multilinear form.

Assume that a family is considering buying a house and there are three houses to consider A, B, and C. Four factors dominate their thinking: house price, remodelling costs, house size as reflected by its footage, and style of the house, which is an intangible. They have looked at three houses with numerical data shown below on the quantifiables (Figure 2).

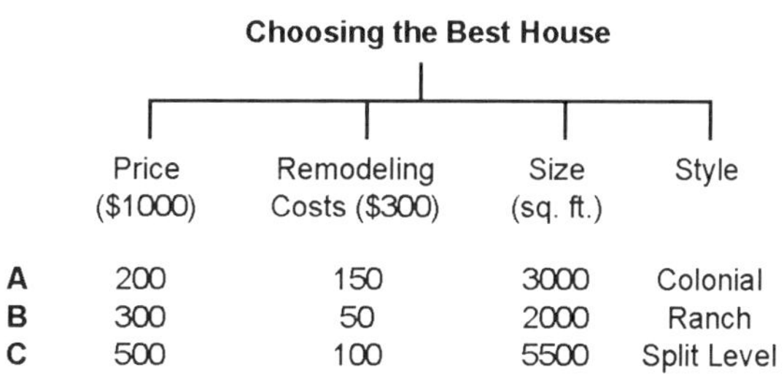

Figure 2. Ranking houses on four criteria.

If we add the costs on price and modelling and normalise we obtain respectively (A,B,C)=(.269,.269,.462). Now let us see what is needed for normalisation to yield the same result. First, we normalise for each of the quantifiable factors. Then we must normalise the factors measured with respect to a single scale (Figure 3).

Choosing the Best House

	Price (1000/1300)	Remodeling Costs (300/1300)	Size (sq. ft.)	Style
A	200/1000	150/300	3000	Colonial
B	300/1000	50/300	2000	Ranch
C	500/1000	100/300	5500	Split Level

Figure 3. Normalising the measurements.

Here we learn two important lessons to be used in the general approach. Normalising the alternatives for the two criteria involving money in terms of the money involved on both criteria leads to relative weights of importance for the criteria. Here for example Price is in the ratio of about three to one when compared with Remodelling Cost and when compared with the latter with respect to the goal of choosing the best house, it is likely to be assigned the value "moderate" which is nearly three times more as indicated by the measurements. Here the criteria Price and Remodelling Cost derive their priorities only from the alternatives because they are equally important factors, although they can also acquire priorities from higher level criteria as to their functional importance with respect to the ease and availability of different amounts of money. We now combine the two factors with a common scale by weighting and adding (Figure 4).

Choosing the Best House

	Economic Factors (combining Price and Remodeling Cost)		Additive Synthesis	Multiplicative Synthesis	Size (sq. ft.)	Style
A	350/1300	=	.269	.256	3000/10500	Colonial
B	350/1300	=	.269	.272	2000/10500	Ranch
C	600/1300	=	.462	.472	5500/10500	Split Level

Figure 4. Combining the two costs through additive or multiplicative syntheses.

The left column and its decimal values in the second column give the exact value of the normalised dollars spent on each house obtained by additive synthesis (weighting and adding). By aggregating the two factor measured with dollars into a single factor, one then makes the decision as to which house to buy by comparing the three criteria as to their importance with respect to the goal.

The second lesson is that when the criteria have different measurements, their importance cannot be determined from the bottom up through measurement of the alternatives, but from the top down, in terms of the goal. The same process of comparison of the criteria with respect to the goal is applied to all criteria if, despite the presence of a physical scale, they are assumed to be measurable on different scales as they might when actual values are unavailable or when it is thought that such measurement does not reflect the relative importance of the alternatives with respect to the given criterion. Imagine that no physical scale of any kind is known! We might note in passing that the outcome of this process of comparison with respect to higher level criteria yields meaningful (not arbitrary) results as noted by two distinguished proponents of multi-attribute value theory (MAVT) Buede and Maxwell (1995), who wrote about their own experiments in decision making:

These experiments demonstrated that the MAVT and AHP techniques, when provided with the same decision outcome data, very often identify the same alternatives as 'best'. The other techniques are noticeably less consistent with MAVT, the fuzzy algorithm being the least consistent.

Multiplicative synthesis, as in the third column of numbers above, done by raising each number in the two columns in the previous table to the power of its criterion measured in the relative total dollars under it, multiplying the two outcomes for each alternative and normalising, does not yield the exact answer obtained by adding dollars! In addition, A and B should have the same value, but they do not with multiplicative synthesis. The multiplicative "solution" devised for the fallacy of always preserving rank and avoiding inconsistency fails, because it violates the most basic of several requirements mentioned in the introduction to this chapter.

Multiplicative and additive syntheses are related analytically through approximation. If we denote by a_i the priority of the ith criterion, i=1,...,n, and by x_i, the priority of alternative x with respect to the ith criterion, then

$$\begin{aligned}\prod x_i^{a_i} = \exp \log \prod x_i^{a_i} &= \exp\left(\sum \log x_i^{a_i}\right) = \exp\left(\sum a_i \log x_i\right)\\ &\approx 1 + \left(\sum a_i \log x_i\right) \qquad (6.3)\\ &\approx 1 + \sum (a_i x_i - a_i) = \sum a_i x_i\end{aligned}$$

If desired, one can include a remainder term to estimate the error. With regard to additive and multiplicative syntheses being close, one may think that in the end it does not matter which one is used, but it does. Saaty and Hu (1998) have shown that despite such closeness on every matrix of consistent judgements in a decision, the synthesised outcomes by the two methods not only lead to different final priorities (which can cause a faulty allocation of resources) but more significantly to *different rankings* of the alternatives. For all these problems, but more significantly because it does not generalise to dependence and feedback even with consistency guaranteed, and because of the additive nature of matrix multiplication needed to compute feedback in network circuits to extend the AHP to the ANP, I do not recommend ever using multiplicative synthesis. It can lead to an undesirable ranking of the alternatives of a decision.

7. RANK PRESERVATION AND REVERSAL

7.1 Theoretical and Practical Issues

Given the assumption that the alternatives of a decision are completely independent of one another, can and should the introduction (deletion) of new (old) alternatives change the rank of some alternatives without introducing new (deleting old) criteria, so that a less preferred alternative becomes most preferred? Incidentally, how one prioritises the criteria and subcriteria is even more important than how one does the alternatives which are themselves composites of criteria. Can rank reverse among the criteria themselves if new criteria are introduced? Why should that not be as critical a concern? The answer is simple. In its original form utility theory assumed that criteria could not be weighted and the only important elements in a decision were the alternatives and their utilities under the various criteria. Today, utility theorists imitate the AHP by rating, and some even by comparing the criteria, somehow. There was no concern then about what would happen to the ranks of the alternatives should the criteria weights themselves change as there were none. The tendency, even today, is to be unconcerned about the theory of rank preservation and reversal among the criteria themselves.

The house example of the previous section teaches us an important lesson. If we add a fourth house to the collection, the priority weights of the criteria Price and Remodelling Cost would change accordingly. Thus the measurements of the alternatives and their number which we call structural factors, always affect the importance of the criteria. When the criteria are incommensurate and their functional priorities are determined in terms of yet higher level criteria or goals, one must still weight such functional importance of the criteria by the structural effect of the alternatives. What is significant in all this is that the importance of the criteria always depends on the measurements of the alternatives. If we assume that the alternatives are measured on a different scale for each criterion, it becomes obvious that normalisation is the instrument that provides the structural effect to update the importance of the criteria in terms of what alternatives there are. Finally, the priorities of the alternatives are weighted by the priorities of the criteria that depend on the measurements of the alternatives. This implies that the overall ranking of any alternative depends on the measurement and number of all the alternatives. To always preserve rank means that the priorities of the criteria should not depend on the measurements of the alternatives but should only derive from their own functional importance with respect to higher goals. This implies that the alternatives should not depend on the measurements of other alternatives. Thus, one way to always preserve rank is to *rate* the alternatives one at a time. In the AHP, this is done through absolute measurement with respect to a complete set of intensity ranges with the largest value intensity value equal to one. It is also possible to preserve rank in relative measurement by using an ideal alternative with full value of one for each criterion.

The logic about what can or should happen to rank when the alternatives *depend* on each other has always been that *anything* can happen. Thus, when the criteria functionally depend on the alternatives, which implies that the alternatives, which of course depend on the criteria, would then depend on the alternatives themselves, rank may be allowed to reverse. The Analytic Network Process (ANP) is the generalisation of the AHP to deal with ranking alternatives when there is functional dependence and feedback of any kind. Even here, one can have a decision problem with dependence among the criteria, but with no dependence of criteria on alternatives and rank may still need to be preserved. The ANP takes care of functional dependence, but if the criteria do not depend on the alternatives, the latter are kept out of the supermatrix and ranked precisely as in a hierarchy (Saaty 1996).

Examples of rank reversal abound in practice, and they do not occur because new criteria are introduced. The requirement that rank always be preserved or that it should be preserved with respect to irrelevant alternatives is not universally accepted. To every rule or generalisation that one may wish to set down about rank, it is possible to find a counterexample that violates that rule. Here is the last and most extreme form of four variants of an attempt to qualify what should happen to rank given by Luce and Raiffa (1957), each of which is followed by a counterexample. They state it but and then reject it. *The addition of new acts to a decision problem under uncertainty never changes old, originally non-optimal acts into optimal ones. The all-or-none feature of the last form may seem a bit too stringent ... a severe criticism is that it yields unreasonable results.* The AHP has a theory and implementation procedures and guidelines for when to preserve rank and when to allow it to reverse. One mode of the AHP allows an irrelevant alternative to cause reversal among the ranks of the original alternatives.

7.2 Selecting the Distributive or Ideal Mode

The distributive mode of the AHP produces preference scores by normalising the performance scores; it takes the performance score received by each alternative and divides it by the sum of performance scores of all alternatives under that criterion. This means that with the Distributive mode the preference for any given alternative would go up if we reduce the performance score of another alternative or remove some alternatives. The Ideal mode compares each performance score to a fixed benchmark such as the performance of the best alternative under that criterion. This means that with the Ideal mode the preference for any given alternative is independent of the performance of other alternatives, except for the alternative selected as a benchmark. Saaty and Vargas (1993) have shown by using simulation, that there are only minor differences produced by the two synthesis modes. This means that the decision should select one or the other if the results diverge beyond a given set of acceptable data.

The following guidelines were developed by Millet and Saaty (1999) to reflect the core differences in translating performance measures to preference measures of alternatives. *The Distributive (dominance) synthesis mode should be used when the decision maker is concerned with the extent to which each alternative dominates all other alternatives under the criterion. The Ideal (performance) synthesis mode should be used when the decision maker is concerned with how well each alternative performs relative to a fixed benchmark.* In order for dominance to be an issue, the decision maker should regard inferior alternatives as relevant even after the ranking process is completed. This suggests a simple test for the use of the Distributive mode: *if the decision maker indicates that the preference for a top ranked alternative under a given criterion would improve if the performance of any lower ranked alternative was adjusted downward, then one should use the Distributive synthesis mode.* To make this test more actionable we can ask the decision maker to imagine the amount of money he or she would be willing to pay for the top ranked alternative. If the decision maker would be willing to pay more for a top ranked alternative after learning that the performance of one of the lower-ranked alternatives was adjusted downward, then the Distributive mode should be used.

Consider selecting a car: Two different decision makers may approach the same problem from two different points of views even if the criteria and standards are the same. The one who is interested in "getting a well performing car" should use the Ideal mode. The one who is interested in "getting a car that stands out" among the alternatives purchased by co-workers or neighbours, should use the Distributive mode.

8. GROUP DECISION MAKING

Here we consider two issues in group decision making. The first is how to aggregate individual judgements, and the second is how to construct a group choice from individual choices.

8.1 How to Aggregate Individual Judgements

Let the function $f(x_1, x_2, \ldots, x_n)$ for synthesising the judgements given by n judges, satisfy the following conditions:

1. *Separability condition* (S): $f(x_1, x_2, \ldots, x_n)=g(x_1)g(x_2)\ldots g(x_n)$ for all $x_1, x_2, \ldots, x_n$ in an interval P of positive numbers, where g is a function mapping P onto a proper interval J and is a continuous, associative and cancellative operation. [(S) means that the influences of the individual judgements can be separated as above.]
2. *Unanimity condition* (U): $f(x, x, ..., x)=x$ for all x in P. [(U) means that if all individuals give the same judgement x, that judgement should also be the synthesised judgement.]
3. *Homogeneity condition* (H): $f(ux_1, ux_2, \ldots, ux_n)=uf(x_1, x_2, \ldots, x_n)$ where $u>0$ and x_k, ux_k (k=1, 2, ..., n) are all in P. [For ratio judgements (H) means that if all individuals judge a ratio u times as large as another ratio, then the synthesised judgement should also be u times as large.]
4. *Power conditions* (P_t): $f(x_1^t, x_2^t, \ldots, x_n^t)=f^t(x_1, x_2, \ldots, x_n)$. [($P_2$), for example, means that if the kth individual judges the length of a side of a square to be x_k, the synthesised judgement on the area of that square will be given by the square of the synthesised judgement on the length of its side.]

Special case ($R=P_{-1}$): $f(1/x_1, 1/x_2, \ldots, 1/x_n)=1/f(x_1, x_2, \ldots, x_n)$. [(R) is of particular importance in ratio judgements. It means that the synthesised value of the reciprocal of the individual judgements should be the reciprocal of the synthesised value of the original judgements.]

Aczél and Saaty (see Saaty 1990 and Saaty 1994) proved the following theorem:

Theorem *The general separable (S) synthesising functions satisfying the unanimity (U) and homogeneity (H) conditions are the geometric mean and the root-mean-power. Moreover, if the reciprocal property (R) is assumed even for a single n-tuple* $(x_1, x_2, \ldots, x_n)$ *of the judgements of n individuals, where not all* x_k *are equal, then only the geometric mean satisfies all the above conditions.*

In any rational consensus, those who know more should, accordingly, influence the consensus more strongly than those who are less knowledgeable. Some people are clearly wiser and more sensible in such matters than others, others may be more powerful and their opinions should be given appropriately greater weight. For such unequal importance of voters, not all g's in (S) are the same function. In place of (S), the weighted separability property (WS) is now: $f(x_1, x_2, \ldots, x_n) = g_1(x_1)g_2(x_2)\ldots g_n(x_n)$. [(WS) implies that not all judging individuals have the same weight when the judgements are synthesised and the different influences are reflected in the different functions $(g_1, g_2, \ldots, g_n)$.]

In this situation, Aczél and Alsina (see Saaty 1994) proved the following theorem:

Theorem *The general weighted-separable (WS) synthesising functions with the unanimity (U) and homogeneity (H) properties are the weighted geometric mean*

$$f(x_1, x_2, \ldots, x_n) = x_1^{q_1} x_2^{q_2} \cdots x_n^{q_n} \tag{8.1}$$

and the weighted root-mean-powers

$$f(x_1, x_2, \ldots, x_n) = \sqrt[\gamma]{q_1 x_1^{\gamma} + q_2 x_2^{\gamma} + \cdots + q_n x_n^{\gamma}} \tag{8.2}$$

where $q_1+q_2+\ldots+q_n=1$, $q_k>0$ $(k=1,2,\ldots,n)$, $\gamma>0$, *but otherwise* $q_1, q_2, \ldots, q_n, \gamma$ *are arbitrary constants.*

If f also has the reciprocal property (R) and for a single set of entries $(x_1, x_2, \ldots, x_n)$ of judgements of n individuals, where not all x_k are equal, then *only the weighted geometric mean* applies. We give the following theorem which is an explicit statement of the synthesis problem that follows from the previous results, and applies to the second and third cases of the deterministic approach:

Theorem *If* $x_1^{(i)},\ldots,x_n^{(i)}$ *i=1,...,m are rankings of n alternatives by m independent judges and if* a_i *is the importance of judge i developed from a hierarchy for evaluating the judges, and hence*

$$\sum_{i=1}^{m} a_i = 1, \tag{8.3}$$

then

$$\left(\prod_{i=1}^{m} x_1^{a_i}\right)^{1/m},\ldots,\left(\prod_{i=1}^{m} x_n^{a_i}\right)^{1/m} \tag{8.4}$$

are the combined ranks of the alternatives for the m judges.

The power or priority of judge i is simply a replication of the judgement of that judge (as if there are as many other judges as indicated by his/her power a_i), which implies multiplying his/her ratio by itself a_i times, and the result follows.

The first requires knowledge how well a particular alternative performs and how well it compares with a standard or benchmark. The second requires comparison with the other alternatives to determine its importance.

8.2 On the Construction of Group Choice from Individual Choices

Given a group of individuals, a set of alternatives (with cardinality greater than 2), and individual ordinal preferences for the alternatives, Arrow proved with his Impossibility Theorem that it is impossible to derive a rational group choice (construct a social choice function that aggregates individual preferences) from ordinal preferences of the individuals that satisfy the following four conditions, i.e., at least one of them is violated:

1. *Decisiveness:* the aggregation procedure must generally produce a group order.
2. *Unanimity:* if all individuals prefer alternative A to alternative B, then the aggregation procedure must produce a group order indicating that the group prefers A to B.

3. *Independence of irrelevant alternatives*: given two sets of alternatives which both include A and B, if all individuals prefer A to B in both sets, then the aggregation procedure must produce a group order indicating that the group, given any of the two sets of alternatives, prefers A to B.
4. *No dictator*: no single individual preferences determine the group order.

The main conclusion about group decision making, using the ratio scale approach of the AHP, is that it can be shown that because now individual preferences are cardinal rather than ordinal, it is possible to derive a rational group choice satisfying the above four conditions. It is possible because: (1) individual priority scales can always be derived from a set of pairwise cardinal preference judgements as long as they form at least a minimal spanning tree in the completely connected graph of the elements being compared and (2) the cardinal preference judgements associated with group choice belong to a ratio scale that represents the relative intensity of the group preferences.

9. CONCLUSIONS

The seven fundamental properties discussed above provide philosophical, mathematical, and practical bases for the AHP and its application. Of primary importance is the capability of the AHP to transform a multidimensional, multi-scale problem into one that is uni-dimenional over a single scale. This allows decision makers to combine vastly different criteria in a rational, context-preserving, and meaningful way. The use of paired comparisons in judgement matrices is intuitively understandable and is easily done in practice. Although calculating priority vectors from these matrices limits the number of elements that can be compared, this difficulty can be easily remedied by absolute rating. In addition, incommensurate element comparisons can be handled by hierarchical clustering that effectively expands the original 1-9 scale to 1-∞. Either rank preservation or rank reversal can be accommodated, depending on the desires of the decision maker and the needs of the decision problem. Finally, cardinal ratio scale preferences permit one to include multiple decision makers in the process and to incorporate their individual judgements in a fair manner that also reconciles their specialised knowledge, experience, and authority.

Any formal decision process (e.g., the AHP, MAVT) tries to capture often ill-formed and complex problems using rational frameworks that appeal to our sense of intelligent decision making. Along the way, assumptions and simplification are made (both implicit and explicit) that make formal decision making practical and manageable. The fundamental properties of the AHP are based on stimulus-response theory, rigorous mathematics, and practical necessities. By doing so, this process mitigates many of the limitations of less "grounded" methods while maintaining broad applicability.

REFERENCES

BUEDE, D. AND MAXWELL, D.T. 1995. Rank disagreement: A comparison of multi-criteria methodologies. Journal of Multi-Criteria Decision Analysis 4: 1-21.

LUCE, R. D. AND RAIFFA, H. 1957. Games and Decisions. Wiley, New York.

MILLET, I. AND SAATY, T. L. 1999. On the relativity of relative measures—accommodating both rank preservation and rank reversal in the AHP. European Journal of Operational Research 121: 205-212.

SAATY, T. L. 1990. Multicriteria Decision Making: The Analytic Hierarchy Process. RWS Publications, Pittsburgh, PA.

SAATY, T. L. 1994. Fundamentals of Decision Making and Priority Theory. RWS Publications, Pittsburgh, PA.

SAATY, T. L. 1996. Decision Making with Dependence and Feedback: The Analytic Network Process. RWS Publications, Pittsburgh, PA.

SAATY, T. L. AND HU, G. 1998. Ranking by eigenvector versus other methods in the Analytic Hierarchy Process. Applied Mathematics Letters 11: 121-125.

SAATY, T. L. AND VARGAS, L. G. 1993. Experiments on rank preservation and reversal in relative measurement. Mathematical and Computer Modeling 17: 13-18.

Chapter 3

On Using the AHP in Multiple Objective Linear Programming

Pekka Korhonen and Jyrki Wallenius
Helsinki School of Economics and Business Administration, Runeberginkatu, Helsinki, Finland

Key words: Multiple objective linear programming, modelling, reference direction

Abstract: We consider how the analytic hierarchy process (AHP) can be used in multiple objective linear programming. In particular, when only qualitative (subjective, judgmental) data is available, the AHP can be used to quantify the qualitative relationships between the row variables and the decision variables. The AHP can also assist a decision maker in specifying the so-called reference direction in an interactive search procedure. The reference direction is a direction, which reflects the desire of the decision maker to improve the values of the objectives starting from the current position. We illustrate our ideas with a numerical example, which is slightly modified from a real application.

1. INTRODUCTION

Since Thomas Saaty developed the analytic hierarchy process (AHP) during the 70's (Saaty 1980), it has become a widely known and used standard method for solving discrete multiple criteria problems. Typically such problems consist of few alternatives and several criteria, possibly having a hierarchical structure. The AHP is a straightforward and transparent method that is also able to consider subjective and judgmental information. These are features that traditionally are missing in multiple objective linear programming (MOLP). In this paper we show that the AHP can be used to advantage in structuring and solving MOLP problems.

Firstly, the AHP can be used in model structuring, when a decision problem—in principle—can be formulated as an MOLP model, except for some qualitative relationships between decision variables and row variables. Such qualitative aspects can be quantified by means of the AHP. Of course,

D.L. Schmoldt et al. (eds.),
The Analytic Hierarchy Process in Natural Resource and Environmental Decision Making, 37–50

this approach is applicable only in some specific models. The original reference is Korhonen and Wallenius (1990).

Secondly, the AHP can be used to support the search procedure in the context of MOLP. The methods developed for solving MOLP problems typically comprise two phases. First the decision maker (DM) is required to give some information concerning his/her preference structure over the multiple objectives and then, using this preference information, the algorithm seeks a solution or a set of solutions for the DM's evaluation. In interactive methods, these phases are repeated until the most preferred solution is found.

Some authors have seen the AHP as a simple and powerful method to obtain preference information from the DM. Kok and Lootsma (1985) have discussed the use of the AHP for finding the weighting vector for the projection function, which is used for projecting an ideal solution onto the efficient frontier. Arbel and Oren (1986) have developed an interactive method for the multiple objective linear programming problem, in which the AHP is used to determine a preference structure over the current solution and its adjacent solutions. Gass (1986) has used the AHP for finding the weights for the deviation variables in goal programming.

In this paper we consider the use of the AHP in finding the so-called reference direction in the visual interactive method developed by Korhonen and Laakso (1986). The reference direction is specified by the DM, and it represents his/her desire to improve the values of the multiple objectives. More details can be found in Korhonen (1987).

In the visual interactive method, each iteration consists of two main steps: determining a search direction and the step-size in this direction. A search direction is found by means of a reference direction. The reference direction can be chosen to be any direction in which the DM's utility is increasing. The reference direction is projected onto the efficient frontier and thus an efficient curve is found for the DM's evaluation.

In their original article, Korhonen and Laakso (1986) used the DM's aspiration levels for specifying a reference direction: the vector from the current solution to the point defined by the DM's aspiration levels is used as a reference direction. Using the aspiration levels, the DM has complete freedom to specify his/her reference direction as he/she likes. Sometimes it may be difficult to find a feasible search direction, in which the values of the objectives are changing in a way similar to the reference direction. To overcome these difficulties, we can try to ask the DM "How would you like to improve the values of the objectives" instead of "In which direction would you like to proceed." It gives more freedom to the system to find a desirable search direction. The AHP is a convenient way to accomplish this.

To solve the step-size problem, the objective values on the efficient curve are shown to the DM using computer graphics in an interactive way and the

DM is asked to choose the best from this set of solutions (see, e.g., Korhonen and Laakso 1986).

This paper consists of five sections. In the second section, we discuss how the AHP can be used to quantify the qualitative relationships between decision variables and row variables. In the third section we review how such models are solved. In the fourth section, we illustrate the use of the AHP to specify a reference direction. An illustrative example is described in the fifth section and concluding remarks are given in the sixth section.

2. USING THE AHP FOR QUANTIFYING A QUALITATIVE RELATIONSHIP

A decision problem may be represented by means of a linear model, if the consequences (outcomes) of the decisions (activities, actions) y_i, $i = 1,\ldots,m$, can be described as linear functions of decision variables $x_j, j = 1,\ldots,n$. Such a model is called a linear decision model, and it may mathematically be expressed as follows:

$$y_i = y_i(\mathbf{x}) = \sum_{j=1}^{n} y_{ij} = \sum_{j=1}^{n} a_{ij}x_j,\ i \in M = \{1, 2, \ldots, m\} \tag{2.1a}$$

or equivalently in the matrix form:

$$\mathbf{y} = \mathbf{y}(\mathbf{x}) = \mathbf{A}\mathbf{x} \tag{2.1b}$$

where $\mathbf{x}$ is an n-vector of decision variables, $\mathbf{A}$ is an $m \times n$ matrix of coefficients, and $\mathbf{y}$ is an m-vector of consequences or outcome variables. The vector $\mathbf{y}$ may include some (or all) of the decision variables x_j if the DM imposes restrictions on (e.g., non-negativity constraints) the decision variables or has preferences over the values of the decision variables.

The problem is to find values for the decision variables x_j, $j \in N = \{1,2,\ldots,n\}$, such that the outcome variables, y_i, $i \in M$, would have desirable values. If $n \geq m$, and $\mathbf{A}$ is of rank m, then for each desired value of $\mathbf{y}$, there exists at least one solution for the model, and it can easily be solved. To avoid this trivial case, we assume that $m > n$.

Depending on the decision problem, model (2.1a,b) may be solved using linear programming, fuzzy linear programming, "what-if"-analysis, or multiple objective linear programming. The basic data needed for each of these models is matrix $\mathbf{A}$. The treatment of the constraints, objectives, etc., is, however, method dependent. In many problems, the elements of $\mathbf{A}$ are easy to find, but often the relationship between a consequence and the decision variables is unknown (qualitative). To treat such a problem by means of a linear decision model, we have somehow to quantify this relationship. We propose the use of the AHP for this purpose.

Many decision problems can be described as linear decision models (2.1a,b). However, some relationships between the decision variables $x_j, j \in N$, and outcome variables y_i, $i \in M$, among other things, may be difficult to specify directly. We note parenthetically that this kind of a decision problem and the corresponding model is called *qualitative*. In the following, we illustrate how these relationships can be subjectively estimated by using the AHP.

We first present each relationship in the difference form:

$$\Delta y_{ij} = c_{ij} \Delta x_j, \ i = 1,2,...,m, \ j = 1,2,...,n \tag{2.2}$$

To begin with, we have to fix the difference in each x_j being considered. It is not necessary that there exist a numerical scale for x_j (monetary or respective). However, often it is possible to describe the change in x_j using soft expressions, such as: "a little", "a lot", "heavy", "much", "somehow", etc. A case in point would be to increase, say, sales promotion "as much as possible" from the current level. Next, one needs to estimate the corresponding change in y_{ij}, independent of whether y_i has a natural numerical scale or not. Using the direct approach, we would ask the DM to respond to a question, such as: "How much will the image of the firm improve, if we increase sales promotion as much as possible?" In case of a numerical scale, this kind of a (direct) question might produce satisfactory results. In the case of a non-numerical scale, it probably does not. Anyway, in either case, the weight assessment technique used in the AHP provides an excellent and systematic way for controlling the estimation error and quantifying a qualitative relationship.

By using the AHP for row i, $i \in M$, we can easily find a vector $\mathbf{w}_i = (w_{i1}, w_{i2}, \ldots, w_{in})$, $\sum w_{ij} = 1$, which describes the relative effects of the change (Δx_j) of each decision variable on the value of row i.

Now, the change Δy_i in row i can be written as follows:

$$\Delta y_i = y_i(\Delta \mathbf{x}) = s_i \mathbf{w}'_i \Delta \mathbf{x}, \ i \in M \tag{2.3}$$

where s_i, $i = 1,2,\ldots,m$, is an (unspecified) scaling factor for the coefficients of row i, $i \in M$, $c_{ij}, j = 1,2,\ldots,n$: $c_{ij} = s_i w_{ij}$.

To find the scaling factors s_i we can use, for example, one of the following principles:

1. $s_i = 1$ or any other constant, $\forall i \in M$,
2. $s_i = 1/\max_j\{w_{ij}\}$, for $i \in M$,
3. s_i is "calibrated" by the DM, e.g., on the basis of a one-unit change in each x_j, $j = 1,2,\ldots,n$, and
4. s_i is calibrated with respect to an "ideal value" of a consequence.

The first principle is appropriate, if the scale of consequence y_i is not very important, and the DM is only interested in how the current value is related to the range of y_i. If each decision variable is allowed to change by one unit, then the change in the value of y_i is equal to one.

The second principle is suitable, when the maximum value/unit has some special meaning for the DM. In a maximization problem this principle implies that a one-unit change in the value of the decision variable with the largest coefficient changes the value of consequence y_i by one unit. In other words, the maximal impact of a one-unit change in the value of a decision variable is one.

If there is a natural scale for some of the rows, then we could calibrate the corresponding outcome variable y_i onto this scale. We may ask the DM to evaluate how large of a change a one-unit change in each decision variable will cause in the outcome variable. This provides us with the following pairs $(\Delta y_{ij}, \Delta x_{ij})$, $j = 1,2,\ldots,n$, in which $\Delta x_j = 1$. We have assumed that $\Delta y_{ij} = s_i w_j \Delta x_{ij}$. The scaling factor can now simply be found through summation:

$$\sum w_j \Delta y_{ij} = s_i \sum w_j \Delta x_j = s_i \sum w_j = s_i . \qquad (2.4)$$

The fourth principle refers to an idea, in which the DM is asked to specify the ideal values (not all zeroes) for the decision variables, and to specify the value of the corresponding outcome. This idea may work for problems, in which the best value for each decision variable is, for example, one and the DM can easily specify the impact of the sum of the variables.

3. "SOLVING" A DECISION MODEL

In this section, we consider the solving of the decision model by using an MOLP approach. An MOLP problem has no unique solution. Any solution on the efficient frontier is acceptable and possible. Which solution will be chosen, depends on the DM's preferences. To obtain preference information from the MD, we may use the AHP for this purpose. In this context, we will discuss how to use the AHP to specify a reference direction in the visual interactive method developed by Korhonen and Laakso (1986). Even if

Korhonen and Wallenius (1988) developed a dynamic version of this method in which there is no need to explicitly specify a reference direction, there may be many problems in which an explicit specification of the reference direction is desirable. For instance, in large-scale problems it is better to have as precise information as possible about the search direction before conducting an actual search to reduce heavy computing time.

We simply assume that "solving" means the following:

$$"max" \, y_i, i = 1,2,...,m \tag{3.1}$$

In practice $m>>0$, this makes it impossible to maximise all consequences simultaneously. That's why we assume that the DM considers the solving of the decision model through *aspiration levels* $b_i \in \Re^m$ for y_i, $i = 1,2,\ldots,m$:

$$y(\mathbf{x}) \geq b_i, i = 1,2,...,m \tag{3.2}$$

The aspiration levels can be called goals. They can further be classified into two classes: flexible and *inflexible (rigid) goals:*

$$\begin{aligned} y_i(\mathbf{x}) &\geq b_i, i \in G \\ y_j(\mathbf{x}) &\geq b_j, j \in R \end{aligned} \tag{3.3}$$

where

- G is the index set of flexible goals and
- R is the index set of inflexible goals.

$M = G \cup R = \{1,2,...,m\}$. We use vector notation $\mathbf{y}^G$ and $\mathbf{y}^R$ to refer to flexible and rigid goal values, respectively. The corresponding notation is used for aspiration level vectors as well.

The flexible goals are only wishes and values exceeding the aspiration levels are preferable as well. Instead, the rigid goal values are crisp and they must be met. Moreover, there are no preferences concerning better values. If there is more than one flexible goal, then we have a multiple objective model, and any solution on the nondominated frontier is a possible solution. Nondominace is defined as follows:

Definition 1. Vector $\mathbf{y}^{G*} \in Y^R = \{\mathbf{y}^R \mid \mathbf{y}^R \geq \mathbf{b}^R\}$ is a *nondominated solution* iff there does not exist another $\mathbf{y}^G \in Y^R$ such that $\mathbf{y}^G \geq \mathbf{y}^{G*}$ and $\mathbf{y}^G \neq \mathbf{y}^{G*}$

Definition 2. Vector $\mathbf{y}^{G*} \in Y^R = \{\mathbf{y}^R \mid \mathbf{y}^R \geq \mathbf{b}^R\}$ is a *weakly nondominated solution* iff there does not exist another $\mathbf{y}^G \in Y^R$ such that $\mathbf{y}^G > \mathbf{y}^{G*}$.

When we refer to the values of the decision variables at a nondominated solution, then the vector $\mathbf{x}^*$ related to a (weakly) nondominated vector $\mathbf{y}^G$ (weakly) is called *(weakly) efficient.* The set of all efficient solutions is called the *efficient set* and the set of all nondominated criterion vectors is called the *nondominated set* (denoted N).

In spite of their different roles, we may present flexible and inflexible goals in a unified way by formulating the following linear programming model (Korhonen and Wallenius, 1988):

$$\begin{aligned} \min \varepsilon \quad & \text{subject to:} \\ & \mathbf{Ax} + \mathbf{w}\varepsilon \geq \mathbf{b} \\ & \mathbf{x} \geq \mathbf{0}, \end{aligned} \tag{3.4}$$

where $\mathbf{w}$ is an m-vector whose components are

$$w_i = \begin{cases} = 0, & \text{if i refers to a constraint row (inflexible goal)} \\ > 0, & \text{if i refers to an objective row (flexible goal)} \end{cases}$$

$\mathbf{b}$ is an m-vector of aspiration levels, and ε is a scalar variable. At the optimum of ε, the solution vector $\mathbf{x}$ is (at least) weakly efficient (Wierzbicki, 1980 and 1986). The formulation (3.4) enables us to consider consequences (that is, objectives and constraints) in a uniform manner.

The (weakly) efficient frontier can be characterized by means of the components $\mathbf{b}^G$ of vector $\mathbf{b}$. Moving on the efficient frontier can be implemented via the following parametric formulation of (3.4) (Korhonen and Laakso, 1986; Korhonen and Wallenius, 1988):

$$\begin{aligned} \min \varepsilon \quad & \text{subject to:} \\ & \mathbf{Ax} + \mathbf{w}\varepsilon - \mathbf{z} \geq \mathbf{b} + t\mathbf{r} \\ & \mathbf{x}, \mathbf{z} \geq \mathbf{0}, \end{aligned} \tag{3.5}$$

where t = 0 initially, and $\mathbf{z}$ is the surplus vector for flexible goals and vector $\mathbf{r}$ is called a *reference direction* and defined as follows:

$$r_i = \begin{cases} = 0, & \text{if } i \text{ refers to a constraint row (inflexible goal)} \\ \neq 0, & \text{if } i \text{ refers to an objective row (flexible goal)} \end{cases}$$

It is used to control the motion on the efficient frontier. By varying the components of $\mathbf{r}$ corresponding to flexible goals, we change the direction of motion. When the DM wants to improve some objectives, we change $\mathbf{r}$

accordingly. In the following, we use vector notation $\mathbf{r}^G$ and $\mathbf{r}^R$ to refer to the components of $\mathbf{r}$ corresponding to flexible and rigid goal values. According to the definition of $\mathbf{r}$, only the components of $\mathbf{r}^G \neq \mathbf{0}$.

4. THE SPECIFICATION OF A REFERENCE DIRECTION BY AHP

The reference direction $\mathbf{r}$ is defined to be a direction at a given solution $\mathbf{y} \in \Re^m$, in which the utility of the DM is at least locally increasing. Because we only consider the flexible goals, we simply write $\mathbf{d} = \mathbf{r}^G$ and $\mathbf{q} = \mathbf{y}^G$. When it is necessary to emphasize the current solution, we mark $\mathbf{q}^0$. Let's assume that the number of flexible goals is p, thus $\mathbf{y}^G \in \Re^p$.

In the above definition, the term "local" means that the DM can take a small step in the direction $\mathbf{d}$ at $\mathbf{q}^0$ (Figure 1), and he/she feels that utility is improving at the moment of evaluation. We do not assume any stable utility function. The utility can be assumed to be changing due to learning and "changes of mind" during the process.

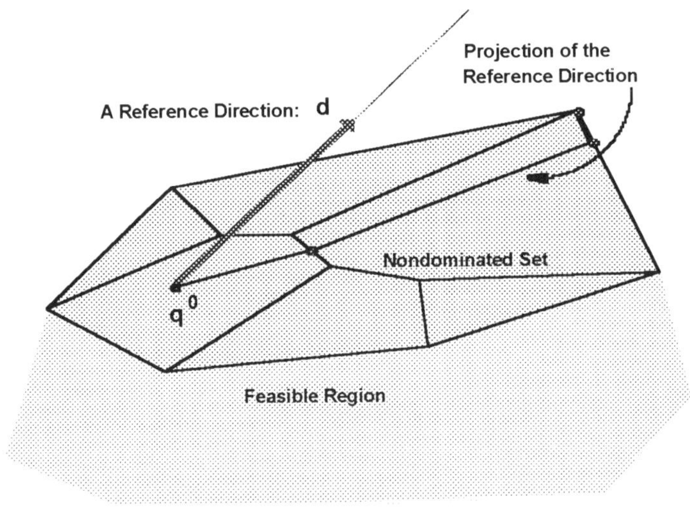

Figure 1. Illustration of the reference direction projection onto the efficient frontier.

The reference direction is easy to find. For example, any direction $\mathbf{d}$, at $\mathbf{q}^0$ is a proper direction. However, $\mathbf{d}$ is not necessarily a feasible direction and thus it is not a possible search direction either. The problem is to find a feasible (= the rigid goals remain fulfilled) search direction, which is somehow related to the reference direction and in which the DM can also find solutions, which are preferable to the current solution. Therefore, the system should help the DM to specify the reference direction in such a way that it is possible to find the projection onto the efficient frontier corresponding to the reference direction. The formulation (3.5) guarantees that the values of the rigid goals $\mathbf{y}^R$ are feasible and the values of the flexible goals $\mathbf{y}^G$ lie on the nondominated frontier.

There are at least three principles to specify a reference direction:

- The DM has complete freedom to specify any reference direction he/she likes.
- The freedom of the DM is partly restricted.
- The DM can make a choice from the set of given (feasible) reference directions.

The original proposal of Korhonen and Laakso (1986) to use the aspiration levels specified by the DM belongs to class 1. The vector starting from the current point and passing through the point of the aspiration levels is used as a reference vector. This direction was projected with model (3.4) onto the (weakly) nondominated frontier, and the resulting (weakly) nondominated curve was presented to the DM in a visual form.

Although the above ideas seem to work quite well, in some problems it is more "economical" to try to have precise preference information from the DM before generating possible solutions. This is the case especially in large-scale models. A promising idea is to use the AHP for this purpose.

By using the AHP, we can find the vector $\mathbf{w} = (w_1,w_2,\ldots,w_n)$, $\sum w_{ij} = 1$which describes the relative importance of improving the values of the objectives at the current point $\mathbf{q}^0$. It has been very interesting to see (Korhonen and Lantto 1986) that without any hesitation people can say "I am more interested in improving the value of objective *i* than objective *j*", although objectives i and j are on completely different scales. Actually, their articulation means "I am more satisfied with the value of objective *j* than that of objective *i*", i.e. they make evaluations on their internal marginal value (utility) scales for objectives.

If we denote $v_i = v_i(q_i)$, $i = 1,2,\ldots,p$, then the above statement may be thought to mean that the people are willing to improve their marginal values $\Delta v_i = v_i(\Delta q_i)$ in such a way that $\Delta v_i > \Delta v_j \geq 0$. Thus we can see the analogy with the AHP philosophy. Comparisons between the changes of their marginal utility values can be likened to comparisons between the weights of stones (Saaty 1980). Using the AHP, we find the vector $\mathbf{w}$, which actually represents the preference structure over the desire of the DM to improve the values of the objectives. If the goal values are not measured in the same scale, the problem is to conclude what is the right interpretation of the sentence "I am more interested in improving the value of objective *i* than objective *j*".

For the reference direction we need the relative changes in terms of the original scales of objectives. Because the transformation rule of the DM from the objective scales into the marginal utility scales is unknown, we assume a simple transformation rule $\Delta v_i = \Delta q_i(q_i^{\max} - q_i^{\min})^{-1}$, where $q_i^{\max}$ and $q_i^{\min}$ refer to the values of objective i, which the DM believes to represent the upper and lower bounds for objective i. The DM is asked to provide these values at the beginning of the process. Using the AHP, we will find the weight vector $\mathbf{w} = (w_1, w_2, \ldots, w_n)$, $\Sigma w_{ij} = 1$, and we write $\Delta v_i = w_i$, $i = 1,2,\ldots p$. The reference direction can be found easily via the following transformation:

$$\Delta q_i = \Delta v_i (q_i^{\max} - q_i^{\min}). \tag{4.1}$$

A similar scaling method for "weights" in goal programming or other approaches has been suggested by many authors (e.g., Kok and Lootsma 1985, Romero *et al.* 1985). In our approach, the "best" and "worst" values for the objectives are specified by the DM, because we believe that his/her internal utility is more related to these values than to values computed by the system.

5. A NUMERICAL EXAMPLE ILLUSTRATING THE USE OF THE AHP

The example used in this section to illustrate the use of the AHP in the MOLP context is slightly modified from the application developed for a small Finnish software company NumPlan Ltd. (Korhonen and Wallenius (1990)). In the late 80's NumPlan Ltd. marketed self-made microcomputer-based decision support systems, whose development and use required expert knowledge. (Therefore, for instance, visibility in the academic world was important to NumPlan Ltd.)

Assume that a company is planning a marketing strategy for itself. The company has introduced the following criteria (objectives/flexible goals) they would like to use in their evaluation (all objectives to be maximized):

- Short run profit (Profit SR).
- Long run profit (Profit LR).
- Scientific Prestige (Prestige).
- Maximum easiness (Easiness).

No numerical information is available. All basic data is obtained from the company board members. Neither demand forecasts nor other types of marketing research information were used in the application.

The following strategies and combinations of strategies were considered:

S1. Sell copies on demand
S2. Direct marketing efforts towards academic colleagues abroad
S3. Direct marketing efforts towards academic colleagues in Finland
S4. Direct advertising to a large population, companies, etc., abroad
S5. Direct advertising to a large population, companies, etc., in Finland
S6. Develop a retailers network
S7. Advertising in management magazines abroad
S8. Advertising in management magazines in Finland
S9. Create publicity through publishing scientific articles describing the methods.

Using the AHP, we first estimate the contribution of each strategy to those four objectives. Scaling coefficients $s_i = 100$, $i = 1,2,3,4$. The results are given in Table 1.

Table 1. The coefficients describing the contribution of each strategy on each objective.

Objective	S1	S2	S3	S4	S5	S6	S7	S8	S9	Min	Max	Range
Profit SR	3	34	10	5	9	14	3	6	16	3	34	31
Profit LR	2	19	3	8	4	36	11	7	10	2	36	34
Prestige	2	17	5	7	3	6	15	4	41	2	41	39
Easiness	33	6	28	3	6	8	3	6	8	3	33	30

The non-shaded portion of Table 1 can be summarized in matrix **A**:

$$\mathbf{A} = \begin{pmatrix} 3 & 34 & \dots & 16 \\ 2 & 19 & \dots & 10 \\ 2 & 17 & \dots & 41 \\ 33 & 6 & \dots & 8 \end{pmatrix}. \tag{5.1}$$

Each strategy can be implemented as a pure strategy or we can develop a mixed strategy by combining pure strategies. To each strategy we associate a decision variable describing how much we will use that strategy in the mixture. Thus each variable varies from 0 to 1 in such a way that 1 stands for a pure strategy. Now we can present our problem as an MOLP-model:

$$\begin{aligned} \max \mathbf{Ax} \quad & \text{Subject to:} \\ & \mathbf{1}'\mathbf{x} \le 1, \\ & \mathbf{x} \ge \mathbf{0}. \end{aligned} \tag{5.2}$$

As discussed in section 3, this model is solved using the formulation:

$$\min \varepsilon \quad \text{Subject to:} \qquad (5.3)$$
$$\mathbf{Ax} + \mathbf{w}\varepsilon \geq \mathbf{b}^G,$$
$$\mathbf{1'x} \leq 1$$
$$\mathbf{x} \geq \mathbf{0},$$

vectors **w** and **b** are p-vectors whose components are

$$\mathbf{w} = \begin{pmatrix} 31 \\ 34 \\ 39 \\ 30 \end{pmatrix} \text{ and } \mathbf{b}^G = \begin{pmatrix} 34 \\ 36 \\ 41 \\ 33 \end{pmatrix}. \qquad (5.4)$$

As initial aspiration levels for the objectives, we use the ideal (maximal) values, and range values (Table 1) are used as the components of the initial weight vector **w**. Solving model (5.3) means that the aspiration levels are projected onto the nondominated frontier. At the optimum, the following decision variables have non-zero values: x_1 = 0.286, x_2 = 0.187, x_6 = 0.230, and x_9 = 0.300 and the values of the objectives are y_1 = 15.18, y_2 = 15.36, y_3 = 17.32, and y_4 = 14.79. The corresponding vector is denoted by $\mathbf{y}^0$.

Assume that the DM would like to find the most preferred direction of improvement. He/she could use the AHP to evaluate the importance of improvement in the objective values. For illustrative purposes we have generated the pair-wise comparison matrix in Table 2.

Table 2. The pairwise comparisons of the objective improvements.

	Profit SR	Profit LR	Prestige	Easiness	Weights
Profit SR	1	1/2	3	7	0.524
Profit LR	2	1	5	9	0.303
Prestige	1/3	1/5	1	5	0.131
Easiness	1/7	1/9	1/5	1	0.041

By multiplying the range values by the components of **w** the DM will find the following reference direction:

$$\mathbf{d} = \begin{pmatrix} 16.24 \\ 10.30 \\ 5.11 \\ 1.23 \end{pmatrix} \qquad (5.5)$$

For instance, by taking a step $t = 0.3$ in the above direction starting from the current solution vector $\mathbf{y}^0$, we will find new aspiration values:

$$\mathbf{b}^{G*} = \begin{pmatrix} 15.18 \\ 15.36 \\ 17.32 \\ 14.79 \end{pmatrix} + 0.3 \begin{pmatrix} 16.24 \\ 10.30 \\ 5.11 \\ 1.23 \end{pmatrix} = \begin{pmatrix} 20.05 \\ 18.45 \\ 18.85 \\ 15.15 \end{pmatrix} \tag{5.6}$$

This solution is found with the variable values: $x_1 = 0.241$, $x_2 = 0.320$, $x_6 = 0.211$, and $x_9 = 0.228$, and the current y values are given in vector $\mathbf{y}^1$:

$$\mathbf{y}^1 = \begin{pmatrix} 18.22 \\ 16.43 \\ 16.55 \\ 13.38 \end{pmatrix} \tag{5.7}$$

If we compare this solution to the previous solution, we see that the DM can obtain the goal values he/she prefers by emphasizing Strategy 2 mainly at the expense of Strategy 9.

6. CONCLUSION

We have described how the AHP can be used in the context of multiple objective linear programming. Firstly, we can help a DM structure the model, when some relationships are purely qualitative, and secondly we may use the AHP to help the DM control the search process in the reference direction approach.

REFERENCES

ARBEL, A. AND OREN, S. 1986. Priority-based interactive multicriteria optimization algorithm. In Preprints of VII-th International Conference on Multiple Criteria Decision Making, Kyoto, Japan, August 18-22, 1986.

GASS S. 1986. A process for determining priorities and weights for large-scale linear goal programmes. Journal of Operational Research Society 37: 779-785.

KOK, M. AND LOOTSMA, F. 1985. Pairwise-comparison methods in multi-objective programming with applications in a long-term energy-planning model. European Journal of Operational Research 22: 44-55.

KORHONEN, P. 1987. The specification of a reference direction using the AHP. Mathematical Modelling 9: 361-368.

KORHONEN, P. AND LAAKSO, J. 1986. A visual interactive method for solving the multiple criteria problem. European Journal of Operational Research 24: 277-287.
KORHONEN, P. AND LANTTO, O. 1986. An experimental comparison of some reference direction techniques for MCDM-problems. Working Paper F-142, Helsinki School of Economics.
KORHONEN, P. AND WALLENIUS, J. 1988. A Pareto race. Naval Research Logistics 35: 615-623.
KORHONEN, P. AND WALLENIUS, J. 1990. Using qualitative data in multiple objective linear programming. European Journal of Operational Research 48: 81-87.
ROMERO, C., SUTCLIFFE, C., BOARD, J. AND CHESHIRE, P. 1985. Naïve weighting in non-preemptive goal programming: Viewpoint and reply. Journal of Operational Research Society 36: 647-649.
SAATY, T. 1980. The Analytic Hierarchy Process. McGraw-Hill, New York.
WIERZBICKI, A. 1980. The use of reference objectives in multiobjective optimization. Pages 468-486 in G. Fandel and T. Gal. (eds) Multiple Criteria Decision Making, Theory and Application. Springer-Verlag, Berlin.
WIERZBICKI, A. 1986. On the completeness and constructiveness of parametric characterizations to vector optimization problems. OR-Spectrum 8: 73-87.

Chapter 4

HERO: Heuristic Optimisation for Multi-Criteria Forestry Decision Analysis

Jyrki Kangas
Finnish Forest Research Institute, Kannus, Finland

Timo Pukkala
University of Joensuu, Joensuu, Finland

Annika S. Kangas
Finnish Forest Research Institute, Kannus, Finland

Key words: Biodiversity, decision support, ecological modelling, forest management planning, heuristics, utility maximisation

Abstract: A heuristic optimisation method (HERO) has been developed for tactical forest planning at the area, or forest holding, levels. In such an approach, the planning area consists of many forest stands, each having several alternative treatment schedules. The idea is to find for each forest stand, or compartment, a treatment schedule that is optimal at the level of the entire planning area. HERO includes both a method for eliciting value judgements and a solution algorithm. It consists of two main phases: estimating a utility model (i.e. analysing and modelling objectives and preferences) and maximising the utility model. Variables in the utility model can be selected from parameters that are associated with the whole forest area, such as drain, costs, income, or qualities of the growing stock. An initial version of HERO used an additive utility model consisting of partial utilities that are determined using so-called sub-priority functions. There are no preconditions on the form of a sub-priority function. Pairwise comparisons among the variables and eigenvalue preference estimation can be used to derive the sub-priority functions and the relative importance of decision criteria. Since 1993, several applications of the initial method have been published. More recent versions of HERO contain extensions, such as multiplicative components in the utility model, which are described in this article.

D.L. Schmoldt et al. (eds.),
The Analytic Hierarchy Process in Natural Resource and Environmental Decision Making, 51–65.

1. INTRODUCTION

There are several requirements that a forest management planning system must satisfy at forest holding or area level. First of all, the system should help to search for the best production program with respect to the objectives set for the forest. In non-industrial private forestry, this means maximising the forest landowner's utility within given restrictions (such as legislation). Second, it should produce treatment alternatives for individual stands or compartments (or other sub-areas) in such a way that certain parameters of the whole forest property attain values that produce the highest possible utility. Third, especially in forestry practice, the management planning system should be easy to use and to understand. Otherwise, it will fail to be accepted in practical forest planning, or its recommendations will not be taken seriously.

Mathematical programming methods and techniques, especially linear programming (LP) and its extension goal programming (GP), are widely applied in research on timber management planning. These methods make it possible to derive treatment alternatives for compartments based on the general production targets for the whole forest. They fulfil the second criterion as listed above. The first criterion, utility maximisation, is fulfilled to some extent, especially by the GP applications. However, both methods assume that utility is linearly related to the goal and constraining variables, or to deviations from the optimal values. For example, the general situation where marginal utility produced by a product is a decreasing function cannot be easily dealt with using LP and GP applications. Also the additivity assumptions of LP and GP are sometimes too limiting from the viewpoint of practical planning. Recently, the limitations of mathematical programming with regard to practical decision situations have been addressed by many researchers, with the aim to develop more realistic forest planning models (e.g. Mendoza and Sprouse 1989, Kangas and Pukkala 1992).

However, limited application of mathematical programming in practical forestry results from the perspective that it is too difficult for most foresters and forest owners to understand and use. People may more readily accept a problem that they cannot solve than a solution that they cannot understand. In these kinds of situations, heuristic methods can be more useful than strict analytical ones. In a heuristic method, the optimal combination of treatments for compartments is found by iterative search methods, and often by interaction between the model and the decision maker, or planner. Heuristic methods are usually easy to understand, and sometimes they can solve complex problems that other methods cannot. Furthermore, heuristics often enable a formulation of the optimisation problem that is more consistent with decision makers' objectives and preferences.

The analytic hierarchy process (AHP) and other multi-attribute decision methods have been applied to forest planning to evaluate management alternatives with respect to preferences of decision makers or of other interested parties (e.g. Hyberg 1987, Mendoza and Sprouse 1989, Kangas 1992). However, forestry planning situations with a great number of decision alternatives, and a long planning period, cannot always be solved using these methods alone. Technical efficiency is needed to handle large optimisation problems. Approaches based on a combined use of multi-attribute decision methods and numerical optimisation techniques seem more promising for analysing complex decision situations.

This paper describes a heuristic method for forest management planning at the forest holding or area level. The method (called HERO) combines the technical efficiency of numerical optimisation and the versatility of multi-attribute decision support to deal with multiple objectives and planning-process-wise preferences. First, HERO is briefly described. Second, an application of HERO to the planning of biodiversity management in boreal forests is presented. Finally, it is shown how multiplicative parts can be added to the utility model as an extension of the initial version of HERO. By using multiplicative utility function components, crisp as well as fuzzy constraints can be dealt with in optimisation. If desired, a purely multiplicative model, or any other utility function, can be used.

2. PRINCIPLES OF HERO

The HERO heuristic optimisation method has been especially developed for tactical forest planning primarily at the area or forest holding level (Pukkala and Kangas 1993). The idea is to seek a combination of stand-level treatment regimes that will provide the best result for the whole area, with respect to the objectives set for forest treatment and utilisation.

Prior to actual optimisation, objectives have to be set for forest treatment and utilisation, and each stand must be assigned a set of alternative treatment regimes for the duration of the planning period (typically 10 to 20 years). Outcomes for alternative regimes are determined using simulation of forest development.

When applying HERO, selecting the best alternative may be divided into two stages: estimation of the utility model and maximisation of this model. Standard HERO uses an additive utility model U, the variables of which are management objectives whose coefficients are the objectives' relative importance (weights), scaled to sum to one (2.1). Objectives are either forest products and values, such as timber, amenity and biodiversity, or resources, such as costs and labour requirements.

$$U = \sum_{i=1}^{n} a_i \cdot u_i(q_i) \tag{2.1}$$

where U is the total utility; n is the number of objectives; a_i is the relative importance of objective i; u_i is the partial utility function, i.e. the sub-priority function of objective i (assuming values in [0, 1], 1 for the best achievable value of the objective measure in question); q_i is the quantity that the plan produces or consumes objective variable i.

In the standard version, the weights a_i are estimated applying pairwise comparisons carried out by the decision maker(s). The relative importance of the objectives can be computed using the eigenvalue method of ratio scale estimation (Saaty 1977). In HERO, the objectives are compared pairwise using a graphical interface, instead of the verbal scale as proposed by Saaty (1980). The relative importance of two objectives is defined by interactively adjusting the lengths of horizontal bars on a computer screen. Practically, the importance of objectives can be determined using other modes of questioning, too.

The relative worth of the planning alternatives with respect to each objective is measured with a sub-priority function. A sub-priority function depicts the change in the utility u_i as a function of the objective variable i. The sub-priority functions scale all objectives to lie between zero and one, making different objectives commensurable.

HERO enables the presentation of objective variables in a hierarchical manner. An objective may be described with a model, the variables of which are the components describing the objective in more detail and the coefficients of which are the components' relative importance. For instance, net income from wood production can be divided into net incomes from different periods, and biodiversity can be estimated from measurements of the components that describe it. This being the case, sub-priority functions are defined to depict the impact of an objective's components on the utility.

When estimating a sub-priority function, first the maximum and minimum values achievable by an objective parameter or component are computed (as single objective optima). In addition, a few intermediate values are also selected. The desirabilities of these values are then estimated by means of pairwise comparisons and the values are given relative priorities, which define a sub-priority function (Figure 1). A sub-priority function can be non-linear, which is often the case, for example, when describing the relationships between biodiversity and environmental variables (e.g., Williams and Gaston 1994). A sub-priority function is estimated separately for each objective measure. Estimation can be based equally well on expertise or subjective value information, or on objective measurements or information produced by empirical research. One advantage of the method is that one operates in real, planning-area-wise production possibilities.

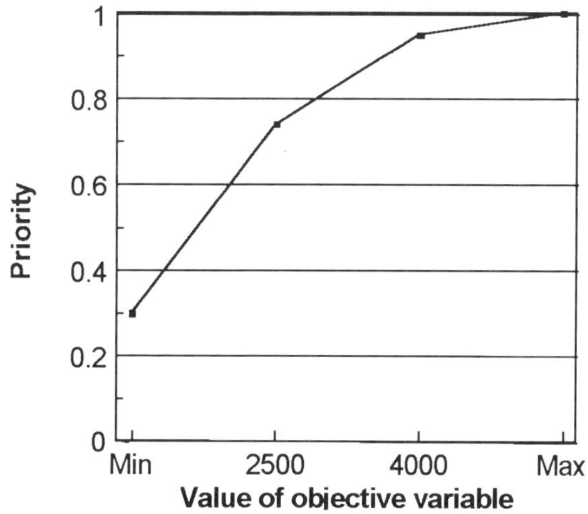

Figure 1. An example of sub-priority functions as applied in HERO.

The sub-priorities are normally scaled so that the maximum sub-priority is always one and other values get sub-priorities relatively to that. If pairwise comparisons and the eigenvalue technique are applied in the estimation of sub-priority functions, results of HERO calculation may differ from corresponding AHP results because of the differences in scaling principles. To avoid this problem, it is possible to use the original AHP priorities as such (without additional scaling) in the estimation of the sub-priority functions. Then, the maximum sub-priorities are not scaled to one. When using HERO, it is essential that the sub-priorities are expressed on a ratio scale. In the estimation of sub-priority functions, other techniques than those used in the AHP could also be applied if they produce relative sub-priority values.

At the "maximisation of overall utility" stage, one uses a heuristic direct-search algorithm to search for the best treatment regime for the forest area. In the beginning of the optimisation process, one treatment schedule is selected randomly for each compartment. The values and the sub-priorities of the objectives are computed, as well as the total utility. Next, one compartment at a time is examined to see whether another treatment regime would increase the utility. If this is the case, the treatment regime that increases utility replaces the current one. Once all the treatment regimes of all the compartments have been revised in this way, the process is repeated again, starting from the first compartment, until no more schedules increasing the utility are found. To ensure that a global optimum is found with an increasing probability, the whole maximisation stage is repeated several times, and the solution with the greatest utility value is taken as the optimal solution.

Also, multiplicative parts or interactive terms can be added to the utility model. This is necessary, if the objective variables have strong interrelationships (e.g., they measure partly the same thing), which is sometimes the case, not only in theory but also in practice. Multiplicative parts are useful when the objective variables are not interchangeable; i.e., if a good gain in one variable cannot compensate for an inferior result in another. Later on in this article, we take a closer look at the use of multiplicative parts in HERO.

When applying HERO, the user has to adhere to objective variables that can be described within the planning system applied. For example, the MONSU software (Pukkala 1993) which makes use of HERO, currently enables the examination of conventional forestry parameters depicting the growing stock (e.g., volumes by tree species) as well as parameters depicting the amount of decaying timber (standing and fallen), mushroom and berry crops, recreational and scenic beauty scores, and indices for biological diversity. So far, applications of HERO include interactive planning of private non-industrial forestry (Kangas *et al.* 1996a), group decision support in forest management planning (Kangas *et al.* 1996b), managing risk and attitudes toward risk in planning calculations (Pukkala and Kangas 1996), incorporating biodiversity into numerical forest planning (Kangas and Pukkala 1996), including variability in forest characteristics at both the stand and area level in calculations (Pukkala *et al.* 1997), and modelling ecological expertise to be used in optimisation (Kangas *et al.* 1998).

3. AN APPLICATION OF HERO: BIODIVERSITY AS A DECISION OBJECTIVE

Implementation of biodiversity for planning calculations can be illustrated by means of a decision hierarchy. Biodiversity is presented as a decision objective in the hierarchy. The components of biodiversity are added into the hierarchy at the level immediately below the level of the objectives. In the same way as the weights of multiple objectives are determined using HERO, the importance of the chosen components of biodiversity are assessed. In an ecosystem-management approach, the set of components consists of environmental variables, which are assumed to indirectly reflect the biodiversity potential of the area in question (within different plan alternatives). Another possibility would be to apply species-wise modelling.

A sub-priority function is estimated for each component. For example, if the volume of dead and decaying wood (m^3/ha) is taken as a component of biodiversity, a sub-priority function is estimated describing the functional relationship between the amount of dead and decaying wood and the related sub-priority.

If needed, more detailed components can be defined with neither theoretical nor methodological problems arising. In that case, sub-priority functions are also estimated for these sub-components. In addition, the importance of the sub-components, with respect to the more general component, should be assessed. For example, the volume of dead and decaying wood might be separated into different tree species. Then, the importance of dead and decaying wood of different tree species, with respect to the considerations of biodiversity, must be assessed, as well as the related sub-priority functions. In this manner, biodiversity can flexibly be implemented in calculations of tactical forest planning (see Kangas and Pukkala 1996, Pukkala *et al.* 1997). The techniques used in the HERO optimisation method allow case-wise choice of biodiversity components as well as their weighting and sub-priority functions. The formula (3.1) can be used to calculate biodiversity indices (BDIs) for forest plans.

$$BDI = \sum_{i=1}^{n} b_i \cdot c_i(q_i) \tag{3.1}$$

where i is a biodiversity component; b is a parameter describing the relative importance of the corresponding component; $c_i(q_i)$ is the sub-priority function describing the contribution of component i to the total biodiversity; others as in the formula above. The greater the index, the better the forest plan in terms of biodiversity considerations. Because of the planning-area-wise calculation procedure, biodiversity indices cannot be universally interpreted nor compared.

When implementing biodiversity for planning calculations, ecological expertise can be utilised. This being the case, the components of biodiversity are chosen, the weights of the components are assessed, and the sub-priority functions are derived on the basis of expert knowledge; i.e. experts on conservation forest biology make the evaluations needed. HERO serves as a framework where expert knowledge can be modelled and integrated into decision support and optimisation.

A case study was carried out in eastern Finland covering about 1500 hectares of state-owned forestland and governed by the Finnish Forest and Park Service. Eleven experts were recruited for the planning process. Before making any comparisons, the experts examined the case study area and its potential for biodiversity management. In the case study, a method for the combined use of HERO and Delphi techniques was developed, with all the comparisons being repeated three times in order to improve the coherence between the judgements of different experts (Kangas *et al.* 1998). Before re-assessing judgements, experts were provided feedback on their previous answers, as well as those of the other experts. In addition to the eigenvalue technique, analyses of pairwise comparisons were made using the variance component modelling approach presented by Alho *et al.* (1996). This method enables statistically sound and versatile analyses of the uncertainty involved in expert predictions.

The mean volume of broadleaf trees, the proportion of old trees, and volume of deadwood were chosen as the components of biodiversity in the case study. These components were regarded to be critical variables with respect to the occurrence of many rare, threatened, and endangered species, which might persist in the planning area. The final model was constructed after the third Delphi round as a mean model of expert views. Sub-priority functions finally accepted in the case study were all non-linear.

In the case study, the resulting formula for calculating biodiversity indices was (Alho and Kangas 1997)

$$BDI = 0.431c_{oldfor}(q_{oldfor}) + 0.258c_{dectre}(q_{dectre}) + 0.311c_{deadwo}(q_{deadwo}) \tag{3.2}$$

where $c_{oldfor}(q_{oldfor})$, $c_{dectre}(q_{dectre})$, and $c_{deadwo}(q_{deadwo})$ are the sub-priority functions of proportion of old forest, volume of deciduous trees, and volume of dead and decaying wood, respectively.

This formula, with non-linear sub-priority functions, can be used in calculating tactical forest planning when HERO optimisation is applied. Decision support information can be generated, for example, on production possibility boundaries for forest biodiversity (at the end of the planning period) in concert with other interesting variables, such as recreation score or removal during the period (Figure 2).

A more detailed model for computing biodiversity indices for alternative forest plans using HERO has been presented by Pukkala *et al.* (1997). In that, an overall biodiversity index consists of two parts calculated on different scales: forest-level diversity and stand-level diversity. A forest-level diversity index was computed from the volume of dead wood, volume of broadleaf trees, area of old forest, and between-stand variation. At the stand level, the area of old forest was replaced by stand age, and variation was described by within-stand variation. All but one of the diversity indicators (stand age) were further partitioned into sub-indicators (Figures 3 and 4). Sub-priority functions were estimated for each indicator at the lowest level.

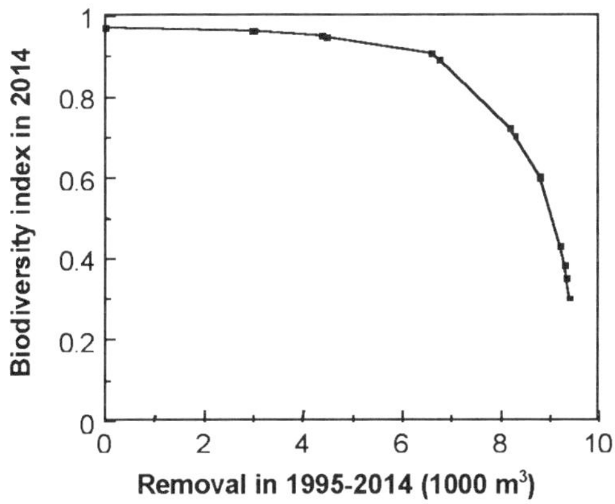

Figure 2. An example of the production possibility boundaries of biodiversity and timber production measures (Kangas and Pukkala 1996).

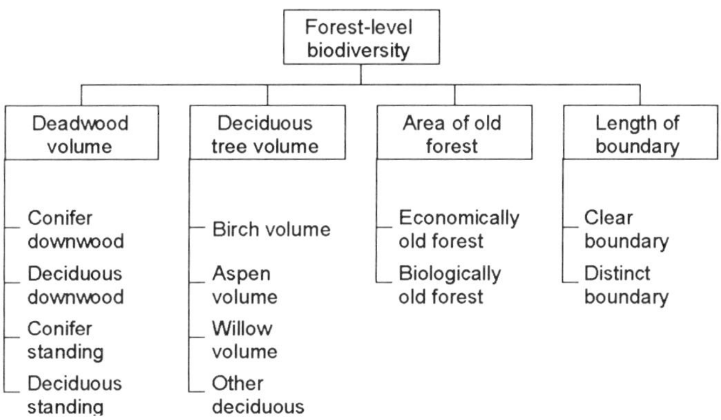

Figure 3. The hierarchy of estimating forest-level species diversity index in a case study in Finland (Pukkala *et al.* 1997).

4. APPLYING MULTIPLICATIVE PARTS IN THE HERO MODEL: THRESHOLD VALUES

When HERO is applied, there are no strict requirements for the form of the utility model used in optimisation. An additive model has been adopted in the standard version mainly because it is easy to estimate, use, interpret, and understand. However, some planning problems cannot be satisfactorily solved using an additive utility function, as already stated above. For example, when applying a species-wise approach in assessing biodiversity matters, if population viability of a species is included in an additive utility function there is no guarantee that viability is greater than zero in the optimal plan. Viability estimates greater than zero can be guaranteed with a multiplicative part in the utility model. Furthermore, in any optimisation problem, strict constraints might be required. Adding a multiplicative part to the model functions similarly to a constraint in mathematical programming, if it is limited to values zero and one only.

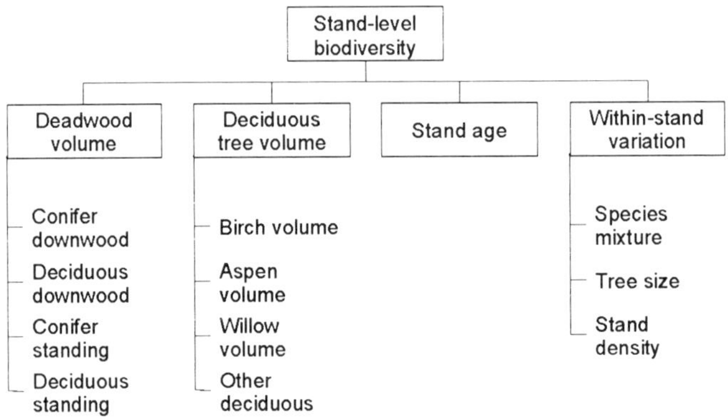

Figure 4. The hierarchy of estimating stand-level species diversity index in a case study in Finland (Pukkala *et al.* 1997).

The general form of the utility model of HERO, when multiplicative parts are included, is (Kangas and Kangas 1998):

$$U = \left(\sum_{i=1}^{n} a_i u_i(q_i) \right) \prod_{j=n+1}^{m} u_j(q_j) \tag{4.1}$$

where the first component is as in (2.1), u_j is the sub-priority function of objective j, q_j is the quantity that the plan produces or consumes objective variable j, m is the total number of objectives including m-n objectives added into the model using multiplicative parts, others as in (2.1).

It would also be possible to apply a purely multiplicative utility model. This being the case, the model is of the form:

$$U = \prod_{i=1}^{n} u_i(q_i) \tag{4.2}$$

As an example of the use of multiplicative parts, we show how species-wise ecological constraints can be included in the utility model. The constraint is described as a threshold value. A given threshold can be obtained by forming a separate function R as the corresponding sub-priority function:

$$R = \begin{cases} 0 & if\ S < S_{crit} \\ 1 & otherwise \end{cases} \tag{4.3}$$

The condition uses a measurable environmental variable S, which is critical for the viability of the species population in question. When function R is used as a multiplicative part in a utility function, it operates similarly to constraints in linear programming, rejecting all solutions not meeting the condition. Thus, an effective way to guarantee population viability in an optimal solution is to form a threshold function, which takes the value of zero if the critical environmental variable is below the threshold, and one if the threshold is exceeded (i.e. the condition is met). If population viability under different forest conditions can be modelled mathematically, the model itself could be used as R.

However, a constraint is seldom completely categorical. Either it is fuzzy by its very nature or uncertainty is involved in determining it. The fuzziness or uncertainty also needs to be accounted for. If, for example, the threshold value guaranteeing population viability is not absolute, the threshold can lie between some minimum and maximum values. When this is the case, using the maximum value of the threshold conservatively maintains viability. Uncertain threshold values for the population size or viability can also be utilised by forming a fuzzy threshold function. A fuzzy threshold function can be defined as (e.g. Mendoza *et al.* 1993):

$$R = \begin{cases} 0 & if\ S_i < S_l \\ \dfrac{S_i - S_l}{S_u - S_l} & if\ S_l \le S_i \le S_u \\ 1 & if\ S_i > S_u \end{cases} \tag{4.4}$$

where S_u is the upper bound of the threshold value and S_l is the lower bound. Thus, an optimal solution will be found among the solutions, which fulfil the minimum threshold value. A fuzzy threshold function may also be defined non-linearly (Figure 5). If (subjective) probabilities of possible threshold values can be assessed, the function R may reflect these probabilities.

5. SOME EXPERIENCES AND CONCLUSIONS

In every forestry-planning problem, both the decision basis and the choice algorithm are needed to find the best—or at least a satisfactory—plan. The decision basis consists of the alternatives available, information about the consequences of alternatives, and the preferences of the decision-maker with respect to these consequences (Bradshaw and Boose 1990).

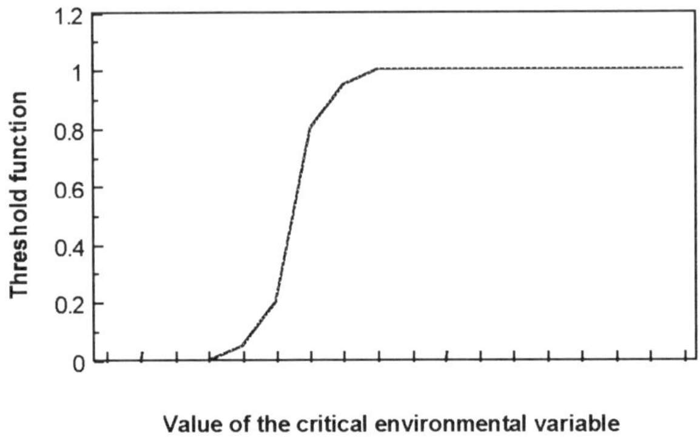

Figure 5. A figure in principle of ecological threshold functions possible to apply as multiplicative parts of the HERO utility model.

In the method presented in this paper, information on alternative management plans, estimation of preferences, and the optimisation process are closely linked to each other. Preferences of the decision-maker are not estimated using hypothetical questionnaires, but with the help of direct quantitatively based questions concerning attainable values of different objective measures. Maximisation of the resulting utility model produces a management plan, or a decision proposal, which yields the most desirable consequences relative to preferences articulated by the decision maker. For planning methods used in forestry practice, it is important that the decision maker feels that articulated preferences really have an effect on the solution. A further advantage of this method is that, using it, not only subjective preferences but also expert judgements can efficiently be utilised in the evaluation of alternative management options. This is often useful, for example, when considering ecological aims, because models based on objective information produced by empirical research may not be available. Our initial experiences, gained from the applications of HERO in tactical forest management planning, have been encouraging. Perhaps the greatest advantage of HERO, from the practical standpoint, is its flexibility—especially with regard to setting objectives and taking individual subjective preferences into account in planning. In that, HERO fits with the idea of value-focused thinking. This is important both in customer-oriented planning of private forestry and in participatory planning. A further advantage is that, due to the sub-priority estimation procedure, specialised expert knowledge is easy to utilise in HERO calculations.

Because HERO does not place limitations on the form of sub-priority functions, it can cope with area-level spatial variables having non-linear utility effects and other non-linearities, more easily than mathematical programming. HERO does not assume that the value of an objective describing a forest area can be calculated as a sum of individual stands, or that the value of an objective in a forest stand does not depend on neighbouring stands. The heuristic method therefore makes it possible to use non-additive objective variables such as variability, biodiversity, and habitat suitability indices for wildlife, among others.

However, some more development and fine-tuning are required before any optimisation method is fully ready for application in routine planning of multiple-use forestry. As we gain more information on various forest uses, objectives, and values, and as planning systems evolve, the range of variables analytically assessable in conjunction with the formulation of plans will grow. This also means that the ways in which different decision criteria and objectives can be integrated will become more diverse and more specific.

Validity of the estimated preferences depends on success in structuring and decomposing the decision problem in question; pairwise comparisons have to be meaningful and easy to understand. In practical forestry, the decision hierarchy needs to be simple and explicit. However, it is always uncertain whether the decision maker is able to give answers that reliably reflect real preferences, even if the estimation method is considered easy. Unfortunately, the validity of a utility model cannot be tested with complete certainty. After all, in forest planning, a utility model provides technical advice only. The aim is to find the best possible treatment program for the forest, not to fully explain or describe preferences. An interactive optimisation process can find good solutions, although the utility model cannot be completely estimated. In that, the "optimal" solution is holistically evaluated after the calculation. If the solution is not regarded the best one or can not be accepted, the utility model is reformulated, the maximisation step is carried out again, etc. The more accurately preferences can be modelled in the first iteration the faster satisfactory plans can be produced. Interactive approaches can be strongly recommended, whatever the optimisation algorithm.

A drawback of any heuristic approach is that the solution may not always be optimal, but only an approximation. Based on tests carried out, and experiences gained in applications, this kind of inefficiency does not seem to be a serious problem: most often the global (technical) optimum is found, and the solutions are always close to optimum. The application of HERO, or any other optimisation method, cannot produce satisfactory solutions to all possible problems of tactical forest management planning. Because intuitive input from decision makers, planners and experts is needed, the planning process is always more or less heuristic, no matter what kind of optimisation algorithms are used.

Concerning the biodiversity application, the assessment of biodiversity still requires improvement: the approaches described above should be seen as preliminary models and as starting points for further development. Neither are applications of the models entirely beyond ecological criticism. For example, the additive assumption as applied in the standard version, is perhaps too restrictive for many components of biodiversity. Also, interactive terms could be added into the utility model. This would, however, make the estimation process more complicated. Instead, multiplicative parts can easily be included in the utility model for biodiversity considerations.

REFERENCES

ALHO, J. AND KANGAS, J. 1997. Analysing uncertainties in experts' opinions of forest plan performance. Forest Science 43:521-528.

ALHO, J., KANGAS, J. AND KOLEHMAINEN, O. 1996. Uncertainty in expert predictions of the ecological consequences of forest plans. Applied Statistics 45:1-14.

BRADSHAW, J. M. AND BOOSE, J. H. 1990. Decision analysis techniques for knowledge acquisition: combining information and preferences using Aquinas and Axotl. International Journal of Man-Machine Studies 32:121-186.

HYBERG, B. T. 1987. Multiattribute decision theory and forest management: A discussion and application. Forest Science 33:835-845.

KANGAS, J. 1992. Multiple-Use Planning of Forest Resources by Uning the Analytic Hierarchy Process. Scandinavian Journal of Forest Research 7:259-268.

KANGAS, A. AND KANGAS, J. 1998. Ekologiset mallit ja ekologisten riskien hallinta metsäsuunnittelussa. Folia Forestalia 2/1998:207-222.

KANGAS, J. AND PUKKALA, T. 1992. A decision theoretic approach applied to goal programming of forest management. Silva Fennica 26:169-176.

KANGAS, J. AND PUKKALA, T. 1996. Operationalization of biological diversity as a decision objective in tactical forest planning. Canadian Journal of Forest Research 26:103-111.

KANGAS, J., PUKKALA, T. AND PYKÄLÄINEN, J. 1996*a*. Vuorovaikutteinen heuristinen optimointi yksityismetsien suunnittelussa. Folia Forestalia 1996(3):231-244.

KANGAS, J., ALHO, J., KOLEHMAINEN, O. AND MONONEN, A. 1998. Analysing consistency of experts' judgments - Case of forest biodiversity. Forest Science 44:610-617.

KANGAS, J., LOIKKANEN, T., PUKKALA, T. AND PYKÄLÄINEN, J. 1996*b*. A participatory approach to tactical forest planning. Acta Forestalia Fennica 251. 24 p.

MENDOZA, G. A. AND SPROUSE, W. 1989. Forest planning and decision making under fuzzy environments: An overview and illustration. Forest Science 35:481-502.

MENDOZA, G. A., BARE, B.B. AND ZHOU, Z. 1993. A fuzzy multiple objective linear programming approach to forest planning under uncertainty. Agricultural Systems 41:257-274.

PUKKALA, T. 1993. Monikäytön suunnitteluohjelmisto MONSU. Ohjelmiston toiminta ja käyttö. Mimeograph. University of Joensuu, Faculty of Forestry. 42 p.

PUKKALA, T. AND KANGAS, J. 1993. A heuristic optimization method for forest planning and decision-making. Scandinavian Journal of Forest Research 8:560-570.

PUKKALA, T. AND KANGAS, J. 1996. A Method for integrating risk and attitude toward risk into forest planning. Forest Science 42:198-205.

PUKKALA, T., KANGAS, J., KNIIVILÄ, M AND TIAINEN, A-M. 1997. Integrating forest-level and compartment-level indices of species diversity with numerical forest planning. Silva Fennica 31(4):417-429.

SAATY, T.L. 1977. A scaling method for priorities in hierarchical structures. Journal of Mathematical Psychology 15:234-281.

SAATY, T. L. 1980. The Analytic Hierarchy Process. McGraw-Hill, New York. 287 p.

WILLIAMS, P. H. AND GASTON, K. J. 1994. Measuring more of biodiversity: Can higher-taxon diversity predict wholesale species richness? Biological Conservation 67: 211-217.

Chapter 5

Strategic and Tactical Planning for Managing National Park Resources

Daniel L. Schmoldt
USDA Forest Service, Southern Research Station, Madison WI USA

David L. Peterson
USGS Forest and Rangeland Ecosystem Science Center, Cascadia Field Station, Seattle WA USA

Key words: Natural resource management, resource allocation, optimisation

Abstract: Each National Park Service unit in the United States produces a resource management plan (RMP) every four years or less. These plans constitute a strategic agenda for a park. Later, tactical plans commit budgets and personnel to specific projects over the planning horizon. Yet, neither planning stage incorporates much quantitative and analytical rigor and is devoid of formal decision-making tools. The analytic hierarchy process (AHP) offers a structure for multi-objective decision making so that decision makers' preferences can be formally accounted for. Preferences for each RMP project, resulting from an AHP exercise, can be used as priorities in an overall RMP. We conducted an exercise on the Olympic National Park (NP) in Washington, selecting eight projects as typical of those considered in RMPs. Five members of the park staff used the AHP to prioritise the eight projects with respect to implicit management objectives. By altering management priorities for the park, three different scenarios were generated. All three contained some similarities in rankings for the eight projects, as well as some differences. Mathematical allocations of money and people differed among these scenarios and differed substantially from what the actual 1990 RMP contains.

1. INTRODUCTION

Resource managers in U.S. national parks must protect a wide array of natural resources, including measurable commodities, aesthetic values, and

D.L. Schmoldt et al. (eds.),
The Analytic Hierarchy Process in Natural Resource and Environmental Decision Making, 67–79.

ecosystem processes (Hinds 1984; Fox *et al.* 1987; Silsbee and Peterson 1991, 1993). Legal and political factors are often at least as important as biological and sociological factors in the development of long-term management plans. Decisions are commonly made in the absence of sufficient technical data or background information. This necessitates the use of expert judgement to evaluate the relative merit of proposed elements of a management plan and to plan for expenditures of time and money.

The selection of resource management activities in national parks is largely driven by how well any activity satisfies overall park management objectives. Projects are combined into a cohesive program to meet large-scale objectives, such as, inventory and monitoring of park resources. In contrast to traditional timber/economic models—e.g., Timber RAM (Navon 1971), MUSYC (Johnson and Jones 1979), and FORPLAN Versions 1 and 2 (Johnson 1986, Johnson *et al.* 1986)—of resource valuation and harvesting, resource management activities in national parks generally are not mutually exclusive and do not necessarily focus on particular tracts of land, e.g. forest stands.

The analytic hierarchy process (AHP) (Saaty 1980) can be applied to resource management decision making to prioritise objectives and alternatives for multi-criteria decisions (Schmoldt *et al.* 1994). This constitutes a *strategic* plan for what park managers would like to accomplish. Such an approach fails, however, to capture some realities, such as multi-year planning and partial allocation that are common in *tactical* planning problems. Therefore, straightforward analytical approaches are needed to allow resource managers to implement management strategies in an optimal manner (e.g., Kangas 1994).

In this paper, we direct our analytical focus on a list of projects included in the current resource management plan (RMP) of a large national park. Specifically, we (1) report on an application of the AHP to prioritise projects in the RMP for Olympic National Park (NP), and (2) compare project priorities that are based on different park management objectives, using Olympic NP as a case study.

2. RESOURCE MANAGEMENT PLANNING

RMPs are required for all National Park Service units in the United States. A standard written format, including budget information, is prescribed and RMPs are reviewed at least every four years. The existing planning process in most national parks is not rigorously structured. The management staff compiles a wide range of topics, discusses them, prioritises them, and develops an RMP with little analysis and without formal decision-making tools. The result is a large and rather cumbersome document, and one that is difficult to justify to others and to modify, as needs change.

As a comprehensive summary of an ideal management strategy, the RMP is a valuable information source, but it is also a source of frustration for park personnel. There is nearly always a huge gap between the management programs described in the RMP and the actual programs that are constrained by budget and personnel limitations. Park managers see many critical needs for information, but they also realise that many of these needs will never be filled. Consequently, they must continually make decisions in the absence of adequate data. They also must choose between an extensive program (many projects at a low level of detail) and an intensive program (a few projects at a high level of detail). Finally, they know that political and operational constraints may override decisions based on scientific information and resource management expertise.

Allocating funds among different resource areas within a national park is a difficult process because of the wide range of resources, personnel, and issues involved. Nevertheless, parks currently have no formal process for prioritising among, and allocating budgets to, projects. The two-step process of prioritisation and allocation presented below introduces analytical rigor into resource management planning. It removes some of the mystery from decision making and allows plans to be re-examined and modified more easily.

3. PRIORITIZING PROJECTS USING THE AHP

3.1 An Overview of the AHP

Many decision-making situations involve preferential selection among alternative items, events, or courses of action. When the selection criterion is "least cost," the measurement scale is obvious and choosing becomes easy. In most real-world situations, however, there is not a single scale for measuring all competing alternatives. More often, there are several scales that must be used and often those scales are related to one another in fairly complex ways.

The AHP (Saaty 1980) helps to structure a problem into a hierarchy consisting of a goal and subordinate features. Subordinate levels of the hierarchy, may include: objectives, scenarios, events, actions, outcomes, and alternatives. Alternatives to be compared—in our case RMP projects—appear at the lowest level of the hierarchy.

3.2 Prioritising Projects in Olympic NP

We used the AHP in co-operation with five Olympic NP staff members to prioritise eight projects (Table 1) out of 147 in the 1990 RMP. The park staff contained highly experienced resource managers with scientific expertise in a wide range of natural-resource disciplines, including vegetation, wildlife, fisheries, and geospatial applications. Using a software implementation of the AHP, pairwise comparisons and project ratings within the AHP were developed interactively by projecting from a computer display directly onto an overhead screen so everyone could discuss the same topic simultaneously. All subjective judgements were reached by consensus of the resource management team through group discussion. In circumstances where consensus cannot be reached easily, separate judgements can be combined by using a geometric average (Schmoldt and Peterson 2000). The following eight projects were selected.

- *Monitor ambient air quality*—Olympic NP is known for its pristine air quality relative to most of the rest of the continental United States. Ambient air is monitored for sulphur dioxide, ozone, total suspended particulates, and visibility.
- *Monitor avalanches*—Subalpine slopes are subject to avalanche hazard in winter, creating problems on developed areas, roads, and ski trails. Avalanche forecasting is critical for visitor safety.
- *Monitor water quality*—Basic physical, chemical, and biological data are needed for water resources throughout the park in order to identify potential changes caused by acidic deposition and human activity.
- *Study and monitor plant communities affected by mountain goats*—Exotic mountain goats potentially threaten plant communities, including some endemic species. Long-term studies are needed to determine if the goats are impacting the growth and distribution of vegetation in alpine and subalpine areas.
- *Conduct studies or management programs for fish or wildlife species of special concern*—There are several threatened, endangered, or sensitive animals in the park, including the northern spotted owl. Populations must be studied to determine their status, and appropriate management actions should be taken if necessary.
- *Inventory and monitor selected anadromous fish stocks that are subject to harvest*—Many fish stocks in the park are managed co-operatively with other agencies and Native Americans. More information is needed on size and distribution of anadromous fish in the park, especially for stocks that have been reduced by harvest and habitat loss.

- *Study and monitor the Elwha watershed*—Two dams on the Elwha River have dramatically changed the aquatic and terrestrial ecosystems in this area. Proposals to remove these dams dictate the need for data on the impact of the presence and subsequent removal of the dams on physical and biological characteristics of the watershed.
- *Conduct an integrated pest management (IPM) program*—Control and eradication of native and non-native species defined as "pests" (wood-rot fungi, carpenter ants, rodents, etc.) are necessary in some developed areas of the park. The use of pesticides and other methods must be monitored and managed responsibly.

Table 1. Priority ratings and rankings for each project under different management objectives.

Project	Objective importance assigned by park staff		All objectives ranked equally		"Management decision making" has highest priority		Actual funding level in the 1990 RMP implicitly determines rankings	
	P[a]	R[b]	P	R	P	R	P	R
Air quality	.137	5	.130	6	.099	7	--	3
Avalanche monitoring	.069	8	.057	8	.111	6	--	2
Water quality	.140	4	.146	3	.122	5	--	5
Goat impacts	.141	3	.135	5	.179	1	--	1
Sensitive wildlife	.143	2	.149	2	.134	4	--	5
Anadromous fish	.128	6	.143	4	.145	3	--	4
Elwha watershed	.148	1	.163	1	.168	2	--	5
IPM program	.095	7	.077	7	.042	8	--	5

[a] Priority value
[b] Ranking

In addition to rating individual projects with respect to each objective and sub-objective, the Olympic NP team developed relative weights for the objectives themselves (Figure 1). Two hypothetical scenarios of objective importance were evaluated for comparison with results from staff-assigned objective priorities. In the first, all objectives were ranked equally—each had the same priority value (Figure 2). For the second scenario, each had a priority value of zero, except for "support management decision making," which had a value of one (Figure 3). For both of these scenarios, the rating scores generated by the park staff for each of the projects across all criteria are the same as above. Different emphases on park management objectives, however, distinguish these scenarios from the original one. These two park management alternatives were chosen because they represent reasonable competing policies for managing park resources.

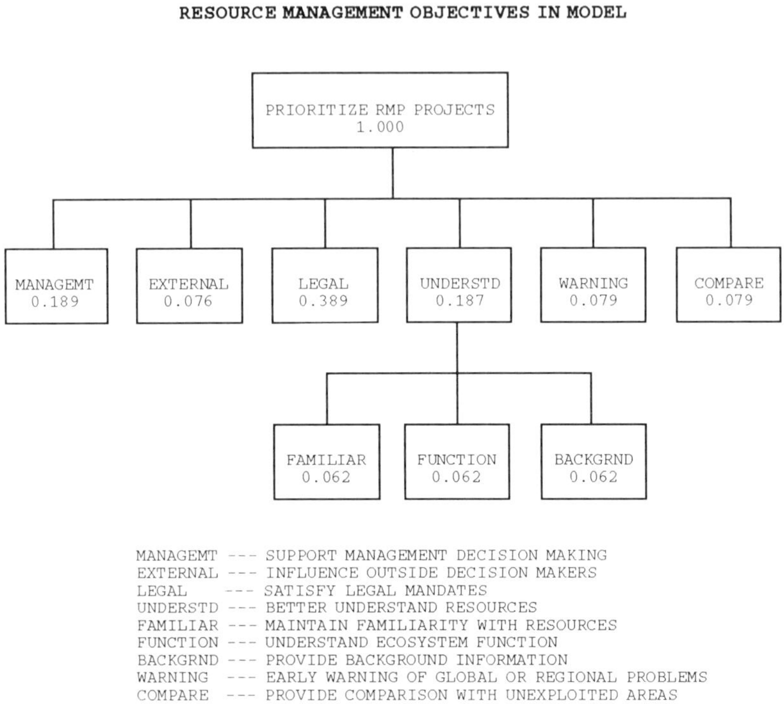

Figure 1. Objectives selected and ranked by park resource management staff are displayed in this hierarchy. Numbers associated with each objective are the global priority values that indicate the importance of each objective for ranking resource management projects

To compare these exercises using the AHP with some real-world results, actual allocation of resources to these eight projects in the 1990 RMP was also used to prioritise them *implicitly* (Table 1). Projects were prioritised based on each project's ratio of allocated to requested expenditures in the actual 1990 RMP. Four unfunded projects out of eight from the 1990 RMP were given an arbitrary ranking of 5 to indicate that they have a lower priority than those ranked 1-4, but otherwise are indistinguishable in rank. We assume here that rankings implied from expenditures provides some insight into implicit priorities by the 1990 RMP decision makers for these projects, i.e. high priority projects would receive a higher percentage of requested expenditures. An exception to this assumption about the expenditure-priority relationship is the avalanche monitoring project; its funding is mandated because it is part of an extensive effort by multiple land management jurisdictions and is relatively inexpensive to implement.

RESOURCE MANAGEMENT OBJECTIVES IN MODEL

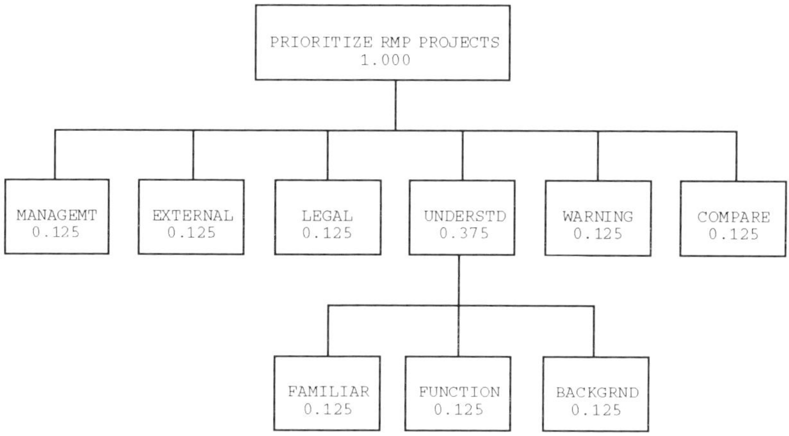

MANAGEMT --- SUPPORT MANAGEMENT DECISION MAKING
EXTERNAL --- INFLUENCE OUTSIDE DECISION MAKERS
LEGAL --- SATISFY LEGAL MANDATES
UNDERSTD --- BETTER UNDERSTAND RESOURCES
FAMILIAR --- MAINTAIN FAMILIARITY WITH RESOURCES
FUNCTION --- UNDERSTAND ECOSYSTEM FUNCTION
BACKGRND --- PROVIDE BACKGROUND INFORMATION
WARNING --- EARLY WARNING OF GLOBAL OR REGIONAL PROBLEMS
COMPARE --- PROVIDE COMPARISON WITH UNEXPLOITED AREAS

Figure 2. All objectives are ranked equally important in this hierarchy. Numbers associated with each objective are the global priority values indicating each objective's importance for ranking resource management projects.

3.3 The Specific Formulation

Because our decision variables X_{ijk} (3.1) are two different types of entities, budget (dollars) and full-time equivalent personnel positions (FTE's), we need some way to put them on the same scale. The conversion factor c_i performs this equilibration of dimensional units. We arbitrarily decided to convert FTE units to budget units; but, without any change in the final solutions, we could have converted budget units to FTE units instead. We then reasoned that the actual allocation of dollars and person-years in the 1990 RMP for these eight projects could be used as a ratio to equate expenditures of budget units and FTE units—an implicit valuation function for budget and FTE's in *this* park and at *this* time. Of the eight projects considered in our example, only four received allocations, which amounted to \$142.6K and 5.2 person-years; therefore, each person-year is equivalent to \$27.4K. Then, because dollars remain unconverted, $c_1 = 1$.

RESOURCE MANAGEMENT OBJECTIVES IN MODEL

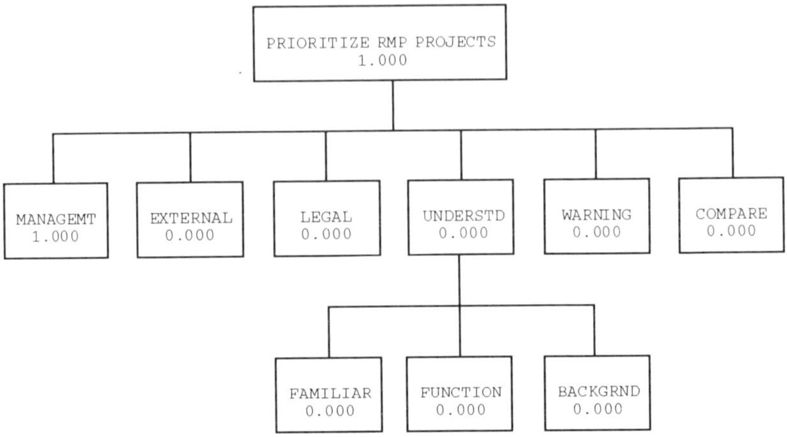

MANAGEMT --- SUPPORT MANAGEMENT DECISION MAKING
EXTERNAL --- INFLUENCE OUTSIDE DECISION MAKERS
LEGAL --- SATISFY LEGAL MANDATES
UNDERSTD --- BETTER UNDERSTAND RESOURCES
FAMILIAR --- MAINTAIN FAMILIARITY WITH RESOURCES
FUNCTION --- UNDERSTAND ECOSYSTEM FUNCTION
BACKGRND --- PROVIDE BACKGROUND INFORMATION
WARNING --- EARLY WARNING OF GLOBAL OR REGIONAL PROBLEMS
COMPARE --- PROVIDE COMPARISON WITH UNEXPLOITED AREAS

Figure 3. Support of management decision making is the only objective in this hierarchy. Numbers associated with each objective are the global priority values indicating the importance of each objective for ranking resource management projects

Our case study example deals with only a small number of projects. The actual 1990 RMP for the park contained 147 projects that were considered for inclusion in the park's management plan. In addition to the constraints listed below (3:2-3.3), several others were added to make the eight-project exercise comparable to the 147-project reality of the actual 1990 RMP (Peterson *et al.* 1994).

Requested expenditures R_{ij} (3.2) and total allocation figures T_j (3.3) were taken directly from the 1990 RMP. Actual allocations for budget and FTE's for all eight projects were assigned to T_1 and T_2. Several additional constraints were included to mirror more closely the implicit allocation methods used in the actual RMP. First, actual 1990 RMP allocations exhibited a nonincreasing flow of expenditures over the four-year planning period. Uncertain future budgets and the problems associated with overly optimistic expectations are a likely reason for this type of planning. This nonincreasing characteristic was reflected in each individual project as well as in the total program. In fact, for each funded project in the 1990 RMP, either all expenditures occurred in the first year or there was an even flow of expenditures over the four years. Because our linear programming software does not allow "exclusive-or" constraints, we used a straightforward nonincreasing inequality. The following set of constraints (3.4) was added to our formulation to reflect these apparent long-term planning realities. An additional set of constraints like those in (3.4), except with "=" replacing "≥", was used for strict even-flow expenditures for "avalanche monitoring."

Objective function:

$$\text{Max } Z = \sum_i \sum_j \sum_k p_i c_j X_{ijk} \tag{3.1}$$

where,

p_i is the priority of project i

c_j is the conversion factor for expenditure type j

X_{ijk} is the expenditure of type j for project i in period k

Subject to:

$$\sum_k X_{ijk} \leq R_{ij} \text{ for } i = 1, \ldots, n \text{ and } j = 1, \ldots, m \tag{3.2}$$

$$\sum_i \sum_k X_{ijk} \leq T_j \text{ for } j = 1, \ldots, m \tag{3.3}$$

where,

n is the number of projects

m is the number of expenditure types

R_{ij} is the total requested expenditure of type j for project i

T_j is the total available expenditure of type j

$$\begin{aligned} X_{ij1} - X_{ij2} &\geq 0 \text{ for } i = 1, \ldots, n \text{ and } j = 1, \ldots, m \\ X_{ij2} - X_{ij3} &\geq 0 \text{ for } i = 1, \ldots, n \text{ and } j = 1, \ldots, m \\ X_{ij3} - X_{ij4} &\geq 0 \text{ for } i = 1, \ldots, n \text{ and } j = 1, \ldots, m \end{aligned} \tag{3.4}$$

Second, not only were expenditures non-increasing, but for the entire RMP, expenditures in the first year amounted to more than 35% of the total expenditures for the four years. Approximately equal budget and FTE units were expended for the subsequent three years of the plan, with greater than

15% of the total for each of those years. We relaxed these actual findings slightly to allow for more latitude in final solutions (3.5).

$$\sum_i X_{ij1} \geq 35\% \text{ of } T_j \text{ for } j = 1, \ldots, m$$
$$\sum_i X_{ijk} \geq 15\% \text{ of } T_j \text{ for } j = 1, \ldots, m \text{ and } k = 2, \ldots, 4 \quad (3.5)$$

Finally, three out of the four projects funded in the 1990 RMP were funded at a level greater than or equal to 18% of requested allocations. The exception was "anadromous fish," which was supported at 8.9% and 5.4% for budget and FTE's, respectively. Projects numbered 1, 2, and 3 in the following constraints (3.6) are the highest ranked projects other than avalanche forecasting. To be consistent with the most conservative allocation from the 1990 RMP, we constrained solutions by requiring that both budget and FTE allocations for each of these three projects be greater than or equal to 5.4% of requested expenditures. A constraint was added to allocate 50% of requested expenditures for "avalanche monitoring" to make our allocation reflect exogenous stipulations used in the 1990 RMP.

$$\sum_k X_{ijk} \geq 5.4\% \text{ of } T_j \text{ for } i = 1, \ldots, 3 \text{ and } j = 1, \ldots, m$$
$$\sum_k X_{[\text{Avalanche forecast}]jk} \geq 50\% \text{ of } T_j \text{ for } j = 1, \ldots, m \quad (3.6)$$

Based on the objective function (3.1) and constraints (3.2-3.6), optimal allocation of budget and FTE units was performed for the different sets of project priorities in Table 1. Results for staff-assigned priorities, for equal objective priorities, for "management decision making" as the only objective, and for the actual 1990 RMP appear in Table 2. To facilitate comparisons with 1990 RMP allocations, at least four projects under each scenario were allocated expenditures as specified in the last constraint (3.6).

4. RESULTS AND CONCLUSIONS

Different scenarios of importance for the objectives in the AHP model produced different project priorities and rankings (Table 1). Projects with the five highest rankings all have relatively high priority scores, while the three lowest priority projects have markedly lower scores. However, a scenario in which "management decision making" is the only objective causes a considerable shift in priorities, such that, "goat impacts" is the highest-ranked project and "anadromous fish" has moved up to third.

Results for the final scenario column, in which rankings are based on the 1990 RMP, differ from each of the previous sets of rankings. Apparently, the park's current, informal process follows a non-explicit set of objectives, which diverges from the explicit objectives of our other scenarios.

Table 2. Based on the LP formulation, optimal total expenditures of budget ($ thousands) and FTE's are displayed for the four scenarios of park objectives

Project	Staff-assigned priorities	All objectives equal	Mgmt decision making	Actual 1990 RMP
Air quality				
Budget				49.6
FTE's				2.0
Avalanche				
Budget	22.00	22.00	22.00	22.00
FTE's	0.60	0.60	0.60	0.60
Water quality				
Budget		20.09		
FTE's		0.20		
Goat impacts				
Budget	6.91		97.72	55.00
FTE's	0.25		3.80	2.2
Sensitive wildlife				
Budget	34.45	34.44		
FTE's	1.33	1.32		
Anadromous fish				
Budget			9.72	16.00
FTE's			0.40	0.40
Sensitive wildlife				
Budget	34.45	34.44		
FTE's	1.33	1.32		
Anadromous fish				
Budget			9.72	16.00
FTE's			0.40	0.40
Elwha watershed				
Budget	79.24	66.04	13.16	
FTE's	3.02	3.09	0.40	
IPM program				
Budget				
FTE's				

When looking at groups of projects, one notices that four out of the five projects, "water quality," "goat impacts," "sensitive wildlife," "anadromous fish," and "Elwha watershed," are the highest ranked projects in each of the first three scenarios. Although some reordering of rankings occurs, these five projects seem to be important regardless of what explicit objectives influence park management.

The LP solutions listed in Table 2 are not unique, but they are optimal. By including more projects (an actual RMP exercise might contain hundreds) or more constraints regarding the relative expenditures between projects or the timing of those expenditures over the planning horizon, it should be possible to create a situation in which there is a unique optimal solution, or even no feasible solution. However, the presence of multiple optimal solutions should not be interpreted negatively, as it provides the park manager with additional latitude to choose a final plan and to react to annual changes in park budgets.

Similarities between priority rankings for the first two scenarios become even more apparent when we examine the allocations listed in Table 2. Except for switching expenditures on "water quality" and "goat impacts," their allocations indicate that they are similar. This suggests that staff-assigned priorities are implementationally most similar (among these scenarios) to treating all objectives as equal. Comparison between scenarios of "staff assigned" priorities and "support management decision making" produces numerous differences. Despite similar project rankings in Table 1, these two scenarios differ substantially in their LP solution. This follows naturally, because the allocation of resources in the LP model is a function of actual priority *values*, and these values may generate very different resource management plans despite similar project *rankings*.

5. DISCUSSION

Most agencies currently have an established structure for developing strategic management plans, but plans are often lengthy and cumbersome, because of efforts to make them comprehensive. Tactical implementation of these plans is generally much less structured. The selection of individual project priorities is rarely quantified, and the rationale for those priorities is not documented. Allocation of limited financial and human resources is often based on criteria that are not quantified or clearly articulated. In general, considerably less effort is devoted to project prioritisation and plan implementation than to the development of the RMP itself.

In the case study conducted for Olympic NP, we found that resource managers are highly receptive to alternative approaches to evaluating a RMP. The complexity of multiple objective planning and project prioritisation was simplified by using the AHP. Furthermore, management staff felt that they could present the RMP to other park staff and the general public with greater confidence if it were based on a more analytical framework grounded in quantifiable decisions. Although this case study assessed only a few projects and objectives, there was considerable support for integrating the AHP approach into more complex aspects of resource management planning.

REFERENCES

FOX, D.G., BERNABO, J. C. AND HOOD, B. 1987. Guidelines for measuring the physical, chemical, and biological condition of wilderness ecosystems. USDA Forest Service General Technical Report RM-146. Rocky Mountain Research Station, Fort Collins, Colorado.

HINDS, W.T. 1984. Towards monitoring of long-term trends in terrestrial ecosystems. Environmental Conservation 11: 11-18.

JOHNSON, K. N. 1986. FORPLAN Version 1: an overview. Land Management Planning System Section, USDA Forest Service, Fort Collins, Colorado.

JOHNSON, K. N. AND JONES, D. B. 1979. A user's guide to Multiple Use-Sustained Yield Resource Scheduling Calculation (MUSYC). Land Management Planning System Section, USDA Forest Service, Fort Collins, Colorado.

JOHNSON, K. N., STUART ,T. W. AND CRIM ,S.A. 1986. FORPLAN Version 2 -- an overview. Land Management Planning System Section, USDA Forest Service, Fort Collins, Colorado.

KANGAS, J. 1994. An approach to public participation in strategic forest planning. Forest Ecology and Management 70: 75-88.

NAVON, D. I. 1971. Timber RAM ... a long-range planning method for commercial timber lands under multiple-use management. USDA Forest Service Research Paper PSW-70.

PETERSON, D. L., SCHMOLDT, D. L. AND SILSBEE, D. G. 1994. A case study of resource management planning with multiple objectives and projects. Environmental Management 18: 729-742.

SAATY, T.L. 1980. The analytic hierarchy process. McGraw-Hill, New York, 287 pp.

SCHMOLDT, D.L. AND PETERSON, D. L. 2000. Analytical group decision making in natural resources: Methodology and application. Forest Science 46: 62-75.

SCHMOLDT, D.L., PETERSON, D. L. AND SILSBEE, D.G. 1994. Developing inventory and monitoring programs based on multiple objectives. Environmental Management 18: 707-727.

SILSBEE, D.G. AND PETERSON, D.L. 1991. Designing and implementing comprehensive long-term inventory and monitoring programs for National Park System lands. Natural Resources Report NRR-91/04. National Park Service, Denver, Colorado.

SILSBEE, D.G. AND PETERSON, D.L. 1993. Planning for implementation of resource monitoring programs. Environmental Monitoring and Assessment 26: 177-185.

Chapter 6

Combined Use of Goal Programming and the Analytic Hierarchy Process in Forest Management

Luis Díaz-Balteiro
Departamento de Ingeniería Agrícola y Forestal. E.T.S. Ingenierías Agrarias. Avda. Madrid, Palencia, Spain

Carlos Romero
Departamento de Economía y Gestión Forestal, E.T.S. Ingenieros de Montes, Avenida Complutense s/n Madrid, Spain

Key words: Decision support, forest management, goal programming

Abstract: This paper presents an analytical framework for forest management taking into account the multiplicity of criteria and decision makers usually present when solving these kinds of decision-making problems. The procedure combines Goal Programming (GP) and the Analytic Hierarchy Process (AHP). In this way, the preferential weights incorporated into the GP model are derived from the application of the AHP method to a group of decision-makers. A key feature of the procedure lies in the ease-of-use and transparent utility interpretation of the solutions obtained. All the theoretical developments were applied to the "Dehesa de la Garganta" forest in the Segovia Mountains ("Sistema Central"), with an area of 2112 hectares covered with Scots pine (*Pinus sylvestris*).

1. INTRODUCTION

The complexity of forest management problems has dramatically increased in recent years because of the multiplicity of purposes and interests involved in this type of decisional context. Nowadays, it is accepted that every decision taken in this field affects several criteria of very different in nature (e.g. economic, environmental, social). It is also accepted that the interests of society as a whole should be pursued in forest management.

D.L. Schmoldt et al. (eds.),
The Analytic Hierarchy Process in Natural Resource and Environmental Decision Making, 81–95.

However, society is made up of different social groups with different perceptions concerning the values of the forests (Bengston and Xu 1995). In short, forest management is in many cases a problem where several criteria as well as several decision-makers are involved.

Forest scientists have adapted and developed methods for dealing with forest-management problems within a multi-criteria framework. Since the pioneering work of Field (1973), involving planning for a woodland property using goal programming (GP), extensive applied literature addressing forest-management problems from the perspective of multiple criteria decision making (MCDM) can be found (e.g. Rustagi and Bare 1987, Mendoza and Sprouse 1989, Liu and Davis 1995). Nevertheless, the task of simultaneous consideration of various criteria and several decision makers is difficult.

This paper reports on an approach combining GP and the analytic hierarchy process (AHP) (Saaty 1980, 1994) and capable of addressing the multiplicity of criteria and decision-makers. Gass (1986) was the first to indicate how weights derived from pairwise comparisons through the AHP could be fruitfully incorporated into a GP model. Kangas and Pukkala (1992) applied the idea to forestry problems. However, the AHP approach is not without its theoretical difficulties (e.g., Dyer 1990, Barzilai and Golany 1994, Barzilai 1998). However, its easy interaction with a decision-maker or a group of decision-makers makes this approach a highly suitable vehicle for deriving preferential weights in different forest management scenarios (e.g. Bing 1988, Peterson *et al.* 1994, Kangas and Pukkala 1996). An important feature of the proposed approach lies in the transparent utility interpretation of obtained solutions.

This paper aims to stimulate the combined use of GP and AHP when tackling forest-management problems. Before presenting the theoretical framework, the main features of a timber-harvest scheduling case study, as well as the main features of the basic model, are briefly described.

2. THE BASIC MODEL: "DEHESA DE LA GARGANTA" FOREST

The "Dehesa de la Garganta" forest in the Segovia Mountains (Spain) is divided into three timber stands: the first two, with a combined area of 379 ha., are managed for conservation purposes while the third, with an area of 2112 ha., consists of Scots pine (*Pinus sylvestris*) plantations managed for commercial wood production. The main characteristics of wood production area are shown in Table 1.

When formulating the harvesting model, the following criteria were considered:

(a) maximisation of the net present value of the forest over the planning horizon;
(b) equality of harvest volume in each cutting period;
(c) area control criterion ultimately aimed at a regulated forest;
(d) final inventory criterion ensuring the perpetuation of wood harvest possibilities. In our case, the initial inventory provided a suitable starting point. Hence, the criterion seeking sustainability of wood harvest leads to an equality between the initial and final inventories and
(e) equality of cash flow in each cutting period.

Table 1. "Dehesa de la Garganta" forest

Age classes	Area			
	Site Class I (ha.)	Site Class II (ha.)	Site Class III (ha.)	Total Area (ha.)
0-20	0	0	0	0
21-40	68	0	16	84
41-60	209	208	184	601
61-80	120	217	405	742
81-100	0	98	326	424
101-120	15	103	136	254
121-140	0	0	0	0
141-160	0	0	7	7
Total	412	626	1074	2112

Source: Prieto & López Quero (1993)

Initially, the model was formulated by considering the net present value criterion as the objective function and the four other criteria as rigid constraints. By taking into account that planning horizon is 100 years, the length time 10 years and the forest rotation on cutting age is between 70 and 120 years, the following linear programming (LP) structure was obtained within a typical Model I formulation (Johnson & Scheurman 1977):

OBJECTIVE FUNCTION:
MAX NET PRESENT VALUE:

$$\sum_h \sum_i \sum_j Nv_{hij} \cdot X_{hij} \tag{2.1}$$

CONSTRAINT SET
Area accounting

$$\sum_j X_{hij} = A_{hi} \qquad \forall\, h, i \tag{2.2}$$

Volume control

$$\sum_h \sum_i V_{hil} \cdot X_{hij} = H_l \tag{2.3}$$

$$-H_l + H_{l+1} = 0 \qquad \forall\, l \tag{2.4}$$

Area control

$$f_K = A \qquad \forall\, K \tag{2.5}$$

Final forest inventory

$$\sum_i V^f_{hiT} \cdot X_{hiT} = I^f_h \qquad \forall\, h \tag{2.6}$$

$$I^f_h = I^i_h \qquad \forall\, h \tag{2.7}$$

Cash flow control

$$\sum_h \sum_i C_{hil} \cdot X_{hij} = L_l \tag{2.8}$$

$$-L_l + L_{l+1} = 0 \qquad \forall\, l \tag{2.9}$$

where:

T = planning horizon.

X_{hij} = hectares of forest harvested for wood from h^{th} class site, i^{th} initial class age at j^{th} prescription.

Nv_{hij} = net present value per hectare of forest harvested for wood from h^{th} class site and i^{th} initial class age at j^{th} prescription

A_{hi} = hectares of forest corresponding to h^{th} class site and i^{th} initial class age

V_{hil} = volume per hectare of forest harvested for wood from h^{th} class site, i^{th} initial class age in l^{th} cutting period.

H_l = volume of wood harvested in l^{th} cutting period.

f_K = hectares of forest belonging to the kth final class age in the final period.

A = forest area divided by the planning horizon in years and multiplied by the time span in years defining the age class.

V^{f}_{hiT} = volume obtained as the result of the final forest inventory on h^{th} site and i^{th} class age.

I^{f}_{h} = volume obtained in final forest inventory on h^{th} site.

I^{i}_{h} = volume obtained in initial forest inventory on h^{th} site.

C_{hil} = cash flow per hectare harvested from h^{th} class site, i^{th} initial class age in l^{th} cutting period

L_l = cash flow obtained in l^{th} cutting period.

Although the linear programming (LP) model (2.1)–(2.9) corresponds to the well-known Model I structure, the following comments are aimed at clarifying its meaning as well as its size.

There are three class sites (h = 1,2,3), six initial class ages (i = 1,2, ...6). The planning horizon is T = 100 years within which ten cutting periods (l = 1, 2, …, 10) with a length of ten years are considered. Moreover, as the time span defining the age class is 20 years, there will be five final age classes (K = 1, 2, ..., 5).

The volume control restraints (2.3) and (2.4) impose strict equality in the wood volume harvested during each of the ten cutting periods considered. Since the time span defining the age class is 20 years, there will be five area-control constraints. Moreover, as the planning horizon is 100 years, area A = (2112 x 20)/(100) = 422 ha. As for the final forest inventory, a constraint ensuring equality between the initial and the final inventory is imposed. Finally, the cash flow constraints (2.8) and (2.9) impose strict equality in the cash flows obtained during each of the ten cutting periods considered. The net present value coefficients Nv_{hij} were calculated using a discount rate of 0.02.

The LP model has a total of 130 decision variables. The area-accounting block represents fourteen constraints; the volume control block represents ten accounting rows plus nine volume constraints; the area-control block represents five constraints (i.e. one constraint for each final class age); the final forest inventory represents three accounting rows (one for each class site) plus three constraints (again one for each class site), and the cash-flow control block represents ten accounting rows plus nine volume constraints. Therefore, the result is an LP model with 130 decision variables and 63 constraints. Readers interested in finding out more details on the overall structure of LP model (2.1)–(2.9) are referred to Díaz-Balteiro (1995).

The above LP problem has no feasible solution. As is usual in this kind of formulation, the feasible set is empty. In other cases, the feasible set, though not empty, may be so small in size that the net present value provided by the model is very low. In short, the LP model (2.1)–(2.9) is too inflexible and should be made flexible in one way or another. In the following sections, the inflexible character of the above problem is mitigated with the help of a GP formulation.

3. A GOAL PROGRAMMING FORMULATION

One way to handle the overly rigid specification of the LP model (2.1)–(2.9) is to treat the right hand side elements as targets that may or may not be achieved. In this way, the constraints are converted into goals. In the transformation process of constraints into goals, only the area-accounting block will keep the character of constraints, whereas the other blocks will be converted into goals. Operating in this way, the following set of goals is obtained:

$$\sum_j X_{hij} = A_{hi} \qquad \forall\, h, i \tag{3.1}$$

$$\sum_h \sum_i \sum_j Nv_{hij} \cdot X_{hij} + n_1 - p_1 = Nv^* \tag{3.2}$$

$$\sum_h \sum_i V_{hil} \cdot X_{hij} = H_l \qquad l = 1, \ldots, 10 \tag{3.3}$$

$$-H_l + H_{l+1} + n_t - p_t = 0 \qquad l = 1, \ldots 9 \qquad t = 2, \ldots 10 \tag{3.4}$$

$$f_K + n_t - p_t = A \qquad K = 1, \ldots 5 \qquad t = 11, \ldots 15 \tag{3.5}$$

$$\sum_i V^f_{hiT} \cdot X_{hiT} = I^f_h \qquad h = 1, 2, 3 \tag{3.6}$$

$$I^f_h + n_t - p_t = I^i_h \qquad h = 1, 2, 3 \qquad t = 16, 17, 18 \tag{3.7}$$

$$\sum_h \sum_i C_{hil} \cdot X_{hij} = L_l \qquad l = 1, \ldots, 10 \tag{3.8}$$

$$-L_l + L_{l+1} + n_t - p_t = 0 \qquad l = 1, \ldots 9 \qquad t = 19, \ldots, 27 \qquad (3.9)$$

where:

n_t = negative deviation variables (t =1,...,27); these variables represent the quantification of the non-achievement of the ith goal.

p_t = positive deviation variables (t =1,...,27); these variables represent the quantification of the over-achievement of the ith goal.

Nv^* = target for the net present value; this target is obtained by solving the following LP model:

$$\begin{aligned} &\text{Max} \sum_h \sum_i \sum_j Nv_{hij} \cdot X_{hij} \\ &\textit{subject to} \sum_j X_{hij} = A_{hi} \qquad \forall\, h, i \end{aligned} \qquad (3.10)$$

In this context, all the deviation variables except p_i are unwanted. The formulation of the GP model implies the minimisation of a certain function of the unwanted deviation variables. The minimisation process can be accomplished by different methods, each one leading to a different GP variant. The two variants considered here are: Archimedean (or weighted GP) and MINMAX (or Chebyshev GP) (see Ignizio 1976, Ignizio and Cavalier 1994). The analytical structure of these formulations appear in (3.11) and (3.12). In both models, the weights w_i reflect preferential as well as normalising purposes. This matter will be clarified in the next section.

From a preferential point of view, the models (3.11) and (3.12) have very different utility interpretations. The Archimedean formulation implies the maximisation of a separable and additive utility function in the criteria considered. This means a solution of maximum efficiency since the sum of the achievements for the criteria considered is maximised. The Chebyshev formulation implies the optimisation of a utility function where the maximum deviation is minimised. This means the most possible balanced solution between the achievements of the different criteria (Ballestero and Romero 1994, Tamiz *et al.* 1998).

a) *Archimedean or Weighted GP Model*
Achievement Function:

$$\begin{aligned} \text{Min} \quad & w_1 n_1 + w_2 \sum_{t=2}^{10} (n_t + p_t) + w_3 \sum_{t=11}^{15} (n_t + p_t) + \\ & w_4 \sum_{t=16}^{18} (n_t + p_t) + w_5 \sum_{t=19}^{27} (n_t + p_t) \end{aligned} \qquad (3.11)$$

Subject to goals and constraints (3.1)–(3.9)

b) *MINMAX or Chebyshev GP Model*
Achievement Function:

$$\begin{aligned}
&\text{Min} \quad D \\
&\textit{subject to: } w_1 n_1 \leq D \\
&w_2 \sum_{t=2}^{10} (n_t + p_t) \leq D \\
&w_3 \sum_{t=11}^{15} (n_t + p_t) \leq D \\
&w_4 \sum_{t=16}^{18} (n_t + p_t) \leq D \\
&w_5 \sum_{t=19}^{27} (n_t + p_t) \leq D
\end{aligned} \tag{3.12}$$

Goals and constraints (3.1)-(3.9)

4. WEIGHTS ELICITATION

As a first step to the implementation of models (3.11) and (3.12), it is necessary to elicit the weights $w_1, \ldots, w_5$ to be attached to the five goals considered. The weights within our context play the following double role:

- They are normalisers of the goals. In our case normalisation is necessary for two reasons: firstly, because the goals are measured in different units (pesetas, hectares and cubic meters of wood), and thus comparing and/or aggregating the unwanted deviation variables is meaningless without prior normalisation; secondly, when the achievement function is optimised, solutions biased towards the goals with higher targets may be obtained.

- They are indicators of the relative preferences of the decision-makers as regards the five goals considered. Thus, the generic weight w_i can be expressed as:

$$w_i = \frac{\alpha_i}{k_i} \tag{4.1}$$

where α_i measures the preferential weight and k_i is the normalising factor for the ith goal.

The normalising function of the weights can be performed by resorting to any normalisation method proposed in the GP field. Here, we adopted weights inversely proportional to the ranges of each goal; i.e., anchor minus anti-ideal values (see Romero 1991, Tamiz *et al.* 1998 concerning the problem of normalising goals). To obtain the anchor and the anti-ideal values for the five goals considered, a pay-off matrix consistent with the GP models formulated was obtained. The entries for each row of the pay-off matrix were obtained by minimising the corresponding unwanted deviation variables and by substituting the optimum vector of decision variables in the other four goals. Operating in this way, the pay-off matrix shown in Table 2 was obtained. The elements of the main diagonal represent the anchor values, and the largest value of each row represents the anti-ideal value for the corresponding goal. Therefore, the following normalising factors are derived from the pay-off matrix:

1. k_1 (Net Present Value)= 740884 - 0 = 740884 thousands of pesetas.
2. k_2 (Volume Control) = 712134 - 0 = 712134 m^3
3. k_3 (Area Control) = 1523 - 0 = 1523 ha
4. k_4 (Final Forest Inventory) = 187920 - 0 = 187920 m^3
5. k_5 (Cash-Flow Control) = 2175345 - 81340 = 2094005 thousands of pesetas.

Table 2. The pay off matrix for the five criteria considered (deviations with respect to target values)

	NPV (thousands of pesetas)	VC (m^3)	AC (ha)	EFI (m^3)	CFC (thousands of pesetas)
NPV	0	551478	546617	585528	740884
VC	712134	0	66750	249072	291987
AC	1523	136	0	845	1084
EFI	54480	21360	98154	0	187920
CFC	2175345	355543	398207	517107	81340

The following highlights should be noted in Table 2. There is a powerful degree of conflict among the five criteria considered. This conflict is especially marked when the net present value objective is optimised. In fact, the wood-harvest schedule of the maximum net present value corresponds to the worst values for volume control, area control, and cash flow criteria. It is also interesting to note that the best outcome in terms of cash-flow control is only compatible with the worst outcomes for net present value and final forest-inventory criteria. Similarly, it should be noted that the cash-flow criterion is the only goal that cannot be fully achieved. It can be said that no solution generated by the single optimisation of any one criterion seems attractive enough to be implemented in practice. Therefore a satisfying solution among the five goals considered should be sought.

The second function of the weights w_i, eliciting the relative preferences of the decision-makers, is approached with the help of the AHP procedure. With this purpose, a dialogue with a group of forestry experts was established as will be shown below.

The hierarchy structure of our problem was as follows. Two aggregate criteria appear in the first hierarchy level: forestry and financial aspects, respectively. The forestry aspect criterion groups three single criteria: volume control, area control, and final forest inventory. The financial aspect groups two single criteria: net present value and cash-flow control. The hierarchy criteria were presented to a group of academic members of staff at the School of Forestry School of the Technical University of Madrid for pairwise comparison. These scholars played the role of an expert committee. The weights shown in Table 3 were then obtained with the help of the EXPERT CHOICE software (Forman *et al.* 1985). It should be noted that only the eight cases analysed or interviews showing an inconsistency index lower than or equal to 0.1 were included.

Table 3. Preferential weights elicited for the hierarchy of criteria considered

Forestry	0.750	0.250	0.750	0.875	0.875	0.500	0.833	0.833
Financial	0.250	0.750	0.250	0.125	0.125	0.500	0.167	0.167
Volume Control	0.281	0.207	0.243	0.078	0.149	0.429	0.091	0.149
Area Control	0.135	0.735	0.088	0.435	0.785	0.143	0.091	0.066
Final Forest Inventory	0.584	0.058	0.669	0.487	0.066	0.429	0.818	0.785
Net Present Value	0.250	0.833	0.250	0.500	0.833	0.250	0.900	0.250
Cash-Flow Control	0.750	0.167	0.750	0.500	0.167	0.750	0.100	0.750

The aggregate preferential weights for the two levels of the hierarchy were determined in two different ways. Firstly, the average values were calculated. Secondly, because of the small sample size (only eight questionnaires), the average values cannot be significant. To redeem this problem, a cluster analysis was implemented and thereby two clusters were obtained. The second cluster comprised the questionnaires corresponding to columns 2 and 5 of Table 3, while the first cluster comprises the questionnaires corresponding to the other six columns of Table 3. The three averages are shown in Table 4.

In order to elicit the final preferential weights, a multiplicative aggregation was implemented between the weights corresponding to the two hierarchical levels shown in Table 3. By resorting to simple multiplication, the final preferential weights shown in Table 5 were obtained.

The weights shown in Table 5 were divided by the corresponding ranges k_i. Thus, the final weight w_1 to be attached to the goal net present value for the cluster 1 was given by:

$$w_1 = \frac{0.0972}{740884} = 0.131 \cdot 10^{-6} \tag{4.2}$$

Table 4. Preferential weights (average cluster 1 and cluster 2)

	Cluster 1	Cluster 2	Average
Forestry	0.757	0.563	0.708
Financial	0.243	0.438	0.292
Volume Control	0.212	0.178	0.203
Area Control	0.160	0.760	0.310
Final Forest Inventory	0.629	0.062	0.487
Net Present Value	0.400	0.832	0.508
Cash Flow Control	0.600	0.167	0.492

Table 5. Final preferential weights for the five criteria considered (average, cluster 1 and cluster 2)

	Cluster 1	Cluster 2	Average
Net Present Value	0.0972	0.3646	0.1483
Volume Control	0.1603	0.1001	0.1437
Area Control	0.1209	0.4275	0.2195
Final Forest Inventory	0.4758	0.0349	0.3448
Cash-Flow Control	0.1458	0.0729	0.1437

Operating in the same way, the following final weights for cluster 1 were obtained for the other four criteria. To avoid scaling problems with the mathematical programming software, all the final weights were multiplied by 10^5.

$$\begin{aligned} w_2 &= 0.225 \cdot 10^{-6} \\ w_3 &= 0.7945 \cdot 10^{-4} \\ w_4 &= 0.253 \cdot 10^{-5} \\ w_5 &= 0.070 \cdot 10^{-6} \end{aligned} \tag{4.3}$$

5. RESULTS AND DISCUSSION

The application of GP models (3.11) and (3.12), to the "Dehesa de la Garganta" forest described in Section 2 and for the weights determined in Section 4, led to the following solutions:

Table 6. Cluster 1

Criteria	*Weighted GP Solution*	*Chebyshev GP Solution*
Net Present Value (thousands of pesetas)	280094	400687
Volume Control (m^3)	229553	236770
Area Control (ha)	39	656
Final Forest Inventory (m^3)	0	20589
Cash-Flow Control (thousands of pesetas)	780053	744133

Table 7. Cluster 2

Criteria	*Weighted GP Solution*	*Chebyshev GP Solution*
Net Present Value (thousands of pesetas)	180762	135109
Volume Control (m^3)	101752	472803
Area Control (ha)	0	237
Final Forest Inventory (m^3)	0	34566
Cash-Flow Control (thousands of pesetas)	583034	1909028

Table 8. Average

Criteria	*Weighted GP Solution*	*Chebyshev GP Solution*
Net Present Value (thousands of pesetas)	470136	231044
Volume Control (m^3)	0	229190
Area Control (ha)	94	321
Final Forest Inventory (m^3)	14991	25210
Cash-Flow Control (thousands of pesetas)	301193	674271

It should be noted here that the above figures represent deviations with respect to the target values. It is interesting to note the perfectly equilibrated character of the Chebyshev solutions. Thus, it is easy to check that the chain of equalities, below, hold for the three cases considered.

That is, the five criteria achieve the same weighted level of fulfilment for the Chebyshev solution. This level ranged from 95.4% (average values) to 93.4% (cluster 2) of their initial targets. It is also easy to check that the level of fulfilment for some of the criteria was highly unbalanced for the weighted GP solution. Thus, the net present value criterion for the average case achieved only the 90.6% of its initial target. In short, the equilibrated character of the Chebyshev solution can be a suitable property for many decision makers striving to avoid feasible schedules excessively biased towards the achievement of one of the criteria.

Cluster 1
$$\begin{aligned} & w_1 \cdot 400687 \\ &= w_2 \cdot 236770 \\ &= w_3 \cdot 656 \\ &= w_4 \cdot 20859 \\ &= w_5 \cdot 744133 \\ &= 0.052 \end{aligned} \tag{5.1}$$

Cluster 2
$$\begin{aligned} & w_1 \cdot 135019 \\ &= w_2 \cdot 472803 \\ &= w_3 \cdot 237 \\ &= w_4 \cdot 34566 \\ &= w_5 \cdot 1909028 \\ &= 0.066 \end{aligned} \tag{5.2}$$

Average
$$\begin{aligned} & w_1 \cdot 231044 \\ &= w_2 \cdot 229190 \\ &= w_3 \cdot 321 \\ &= w_4 \cdot 25210 \\ &= w_5 \cdot 674271 \\ &= 0.046 \end{aligned} \tag{5.3}$$

It was observed in Section 4 that no solution generated by the single optimisation of any criterion seemed attractive (Table 2). On the contrary, the satisfying solutions generated by our GP models seem attractive and feasible in practice. However, it should be noted that the weighted and Chebyshev solutions vary greatly between the two clusters and the average values. This result is easy to understand given the extremely different set of weights attached to each of the three groups considered. In fact, each group (clusters 1 and 2, and the average values) represent different subjective orientations toward the importance of each criterion in forestry planning.

6. CONCLUDING REMARKS

This paper supports previous findings (Díaz-Balteiro and Romero 1997, 1998) suggesting that integrating several MCDM methods (Compromise Programming, GP, AHP, etc.) can help address forest-management problems where several criteria are considered by several groups of decision makers.

The methodological approach proposed seems attractive at least for the following reasons. First, it accommodates the multiplicity of criteria involved in any forest-management problem. Second, the solutions generated by the model can be easily interpreted in utility terms. Thirdly, it is relatively easy to interact with a decision-maker or groups of decision-makers in order to derive the weights reflecting the corresponding group preferences.

ACKNOWLEDGEMENT

This research was supported by "Comisión Interministerial de Ciencia y Tecnología (CICYT)" and "Consejería de Educación y Cultura de Madrid." Christine Mendez and Erkki Pekkinen revised the English language.

REFERENCES

BALLESTERO ,E. AND ROMERO, C. 1994. Utility optimisation when the utility function is virtually unknown. Theory and Decision 36:233-243.

BARZILAI, J. 1998. On the decomposition of value functions. Operations Research Letters 22:159-170.

BARZILAI, J. AND GOLANY, B. 1994. AHP rank reversal, normalisation and aggregation rules. INFOR 32:57-64.

BENGSTON, D. N. AND XU, Z. 1995. Changing national forest values: A content analysis. United States Department of Agriculture. North Central Forest Experiment Station, Research Paper NC-323.

BING, N. G. 1988. A simple algorithm of the dynamic priorities and an approach to Chinas forestry production structure for the future, in Proceedings of International Symposium on the Analytic Hierarchy Process, Tianjun University, China, 461-464.

DIAZ-BALTEIRO, L. 1995. Modelos de programación matemática para la ordenación de montes: desarrollos teóricos y aplicaciones al sector forestal español. Doctoral dis, Forestry School, Technical University of Madrid.

DÍAZ-BALTEIRO, L. AND ROMERO, C. 1997. Timber harvest scheduling problems: compromise programming and utility optimization models. Lecture Notes in Economics And Mathematical Systems 455:328-337.

DÍAZ-BALTEIRO, L. AND ROMERO, C. 1998. Modeling timber harvest scheduling problems with multiple criteria: an application in Spain. Forest Science 44:37-46.

DYER J. S. 1990. Remarks on the analytic hierarchy process. Management Science 36: 249-258.

FIELD, D.B. 1973. Goal programming for forest management. Forest Science 19:125-135.

FORMAN, E.H., SAATY, T.L., SELLY, M.L. AND WALDROW, R. 1985. EXPERT CHOICE, Decision Support Software Inc., Pittsburgh.

GASS, S. I. 1986. A process for determining priorities weights for large-scale linear goal programmes. Journal of the Operational Research Society 37:779-785.

IGNIZIO, J.P. 1976. Goal programming and extensions. Lexington Books, Massachusetts.

IGNIZIO J.P. AND CAVALIER, T. M. 1994. Linear programming. Prentice-Hall, New Jersey.

JOHNSON, K.N. AND SCHEURMAN, H.L. 1977. Techniques for prescribing optimal harvest and investment under different objectives. Discussion and synthesis. Forest Science Monographies 18.

KANGAS, J. AND PUKKALA, T. 1992. A decision theoretic approach applied to goal programming of forest management. Silva Fennica 26:169-176.

KANGAS, J. AND PUKKALA,T. 1996. Operationalization of biological diversity as a decision objective in tactical forest planning. Canadian Journal of Forest Research 26:103-111.

LIU, G. AND DAVIS, L.S. 1995. Interactive resolution of multi-objective forest planning problems with shadow price and parametric analysis. Forest Science 41:452-469.

MENDOZA, G.A. AND SPROUSE, W. 1989. Forest planning decision making under fuzzy environments: an overview and illustration. Forest Science 35:481-502.

PETERSON, D.L., SILSBEE, D.G. AND SCHMOLDT, D.L. 1994. A case study of resources management planning with multiple objectives and projects. Environmental Management 18:729-742.

PRIETO, A. AND LÓPEZ-QUERO, M. 1993. Manual de ordenación de montes. Editorial Paraninfo, Madrid.

ROMERO, C. 1991. Handbook of critical issues in goal programming. Pergamon, Oxford.

RUSTAGI, K.P. AND BARE, B.B. 1987. Resolving multiple goal conflicts with interactive goal programming. Canadian Journal of Forest Research 17:1401-1407.

SAATY, T.L. 1980. The analytic hierarchy process: planning, priority setting, and resource allocation. McGraw-Hill, New York.

SAATY, T.L. 1994. How to make a decision: the analytic hierarchy process. Interfaces 24:19-43.

TAMIZ, M., JONES, D. AND ROMERO, C. 1998. Goal programming for decision making: An overview of the current state-of-the-art. European Journal of Operational Research 111:569-581.

Chapter 7

Efficient Group Decision Making in Workshop Settings

Daniel L. Schmoldt
USDA Forest Service, Southern Research Station, Madison WI USA

David L. Peterson
USGS Forest and Rangeland Ecosystem Science Center, Cascadia Field Station, Seattle WA USA

Key words: Group decision making, workshops, brainstorming, fire research

Abstract: Public land managers must treat multiple values coincidentally in time and space, which requires the participation of multiple resource specialists and consideration of diverse clientele interests in the decision process. This implies decision making that includes multiple participants, both internally and externally. Decades of social science research on decision making by groups have provided insights into the impediments to effective group processes. Nevertheless, there has been little progress in producing more rigorous and accountable decision processes in land management. The authors' experiences with temporary, formal groups (workshops) have led them to develop a process for group decision making that combines (1) a strawman document to initiate and pattern group discussion, (2) brainstorming to generate ideas, and (3) the analytic hierarchy process to produce judgements, manage conflict, and develop implementation plans. An application of this group process to program development in fire research in a workshop setting indicates that it is efficient and cost effective, and provides a large amount of useful quantitative information about group preferences.

1. INTRODUCTION

Natural resource management has become increasingly complex during the past two decades due to the multiplicity of management objectives that must be considered to address public interest, legislative requirements, and

D.L. Schmoldt et al. (eds.),
The Analytic Hierarchy Process in Natural Resource and Environmental Decision Making, 97–114.

environmental compliance. "Ecosystem management" is the paradigm most commonly cited as the appropriate template for resource management by public agencies. Indeed, this concept has provided a vehicle for a transition on United States federal lands from commodity-dominated and output-based management to the inclusion of multiple resource values.

Few choices in natural resource or environmental management are made unilaterally. Decision makers rely on others either directly through consultation and collaboration or indirectly through established protocols and chains of command. There is a tacit belief that groups function in a superior way to individuals when important issues are at stake, which has led to a proliferation of workshops focused on a wide range of issues in natural resources. While there are many important benefits from group interaction and a team approach to problem solving, there are also well-documented drawbacks associated with group processes (McGrath 1984). In light of the growing complexity of decisions in natural resource management, group decision making is becoming increasingly common, and we anticipate that its shortcomings will become more noticeable in the future.

Many decisions that must be made depend on subjective information and values. Judgmental (value laden) decisions that do not result in group unanimity produce less decision satisfaction for group members (Kaplan and Miller 1987), as opposed to informational (intellective) decisions that have a demonstrably "correct" answer. This implies that as strategic and tactical land management decisions are influenced by a wide variety of stakeholders' agendas (not entirely intellective influences), it will become more difficult for a majority to reach a state of satisfied acceptance. Therefore, it is increasingly important that differences in preferences be understood and that mechanisms and procedures for describing and handling them be developed and applied.

Many natural resource problems involve selecting among a fixed set of alternatives or treatments or scenarios—a 1-of-*N* decision situation. On the surface, this seems like a straightforward task, but it is not that simple. There are many criteria, influences, and stakeholders that help to frame a decision. This often reduces the likelihood of making a good decision to little better than 1/*N*, or random odds.

Furthermore, decision making typically involves a BOGSAT process ("Bunch Of Guys/Gals Sitting Around a Table", Peterson *et al.* 1994). BOGSAT appears, on the surface, as a very cost-effective decision mechanism, because relatively little time or effort is expended. These perceived cost savings can become irrelevant, however, if shortcomings of the process lead to downstream costs such as time-consuming and expensive litigation and land *mis*management. By expending more organised and systematic effort up-front, it may be possible to reduce total costs in terms of time, money, and credibility.

Because we expect that dependence on group decision making (GDM) in natural resource management will increase, we have sought to develop a group decision process that minimises negative dynamics and process losses, while attaining beneficial group effects. Based on a review of the pertinent social science literature and our own empirical observations during group decision-making situations, we have developed a group process that contains three basic components: (1) a strawman document that acts as a template and starting point for group discussion, (2) a mechanism for idea generation that enables a group to quickly and easily produce issues to be included in the decision process (e.g., criteria, objectives, alternatives, etc.), and (3) the analytic hierarchy process (AHP) as the decision structuring and analysis component. In the next section, we provide some background and previous research results on GDM, followed by a description of our GDM approach for workshops and other formal meetings.

2. GROUP DECISION MAKING

2.1 Group Attributes and Tasks

In some instances, decision-making groups contain relatively fixed membership and persist for long periods of time, meeting periodically to make strategic, policy, or tactical decisions (e.g., the resource management staff of a national forest—a persistent, formal group). Other groups are assembled for a short period of time for specific tasks (e.g., technical workshops—temporary, formal groups, q.v. Peterson *et al.* 1992, Rogelberg *et al.* 1992, Peterson *et al.* 1993, Schmoldt *et al.* 1999). Such task-oriented, temporary groups can be distinguished by differentiation of members' skills, little synchrony within or across members' organisations, and variable duration (Sundstrom *et al.* 1990). While these two types of groups (and specific groups, as well) may differ in decision rules, group dynamics, membership, meeting procedures, and organisational support, all types of groups have common problems (see Group Liabilities, below).

It is often assumed that decisions produced by a group are superior to decisions by an individual. In reality, groups generally perform better than their *average* individual member does but worse than the group's *best* individual (Hall and Watson 1970, Hill 1982, Yetton and Bottger 1982, Bottger and Yetton 1987, Rogelberg at al. 1992). Ideally, we should strive to avoid group deficiencies and yet capitalise on inherent group benefits. All types of groups can benefit from group-decision methods that facilitate dialog, mitigate adverse interactions, provide a smooth and efficient process, and produce good collective decisions.

McGrath (1984) summarised much of the existing literature on group interaction and performance, and categorised group tasks into four components: (1) generate (identify alternatives), (2) choose (make value-laden judgements), (3) negotiate (manage conflict), and (4) execute (coordinate detailed implementations). Most resource management decisions and actions incorporate aspects of each of these dimensions, which makes analysis and implementation difficult.

A group-decision context provides several benefits. First, two individuals bring more knowledge to the table than one person does; each additional person brings an added amount. Second, the addition of other people to the decision process also produces an interaction effect, whereby multiple approaches to a problem can eliminate the limited scope that often hinders individual thinking. Third, if more than one person is affected by a decision, it is desirable to have those affected parties involved in the decision process. Participation increases decision acceptance and the ability and willingness of group members to champion the decision when faced with affected parties outside of the group. Because these assets are intrinsic to most groups, most research has sought to identify which factors *hinder* GDM, and to find methods that eliminate them.

2.2 Group Liabilities

"Process losses" (Steiner 1972) associated with human interaction impede group communication. On the other hand, when group interaction favours the exchange of relevant decision-making information, favourable decision outcomes occur (Vinokur *et al.* 1985). Shyness, poor communication skills, and individual dominance all contribute to process losses in groups (Johnson and Johnson 1987). Social pressures to conform can stifle effective discussion (Maier 1967) and lead to group avoidance of viable alternatives (groupthink). Social loafing—relying on others to perform the group's work—is also common (Williams *et al.* 1981). Additional problems include personality conflicts (Maier 1967), promotion of personal agendas, and uncooperative individuals.

Agreement within a group (consensus) is important because it: (1) ensures individual ownership in, and commitment to, the group solution, (2) promotes individual satisfaction with the group outcome, (3) provides a unified (even if only majority) group decision that is viewed as more reliable and supportable by outside agents, and (4) produces a group accomplishment and avoids the perception of a lack of consensus. Majority and unanimity are the two basic decision rules used to obtain consensus (conformity in the case of majority rule). On the other hand, expectations to conform and produce a consensus judgement can often dilute individual, specialised contributions. The failure by groups to adequately consider and accept individual opinion (when correct) often drives suboptimal group performance (Maier and Solem 1952, Janis 1971, Lamm and Trommsdorff 1973). Consequently, groups often choose a middle-ground position that compromises a better alternative for the sake of agreement (cohesion; Callaway and Esser 1984, Leanna 1985) or to merely avoid a less desirable alternative.

The authors' experiences with technical workshops (as temporary formal groups) suggest that such meetings often are dominated by unfocused and rambling discussion, which mixes judgmental and intellective issues (Schmoldt and Peterson 1991, Peterson *et al.* 1992, Peterson *et al.* 1993, Schmoldt *et al.* 1999). Ideas presented in such a freeform dialog have merit, but those ideas may not always be synchronised with a logical flow of topics. While general discussions of this nature can produce beneficial results due to juxtaposed ideas, there is also a cost due to inefficiencies of time and effort and the potential loss of ideas introduced in the wrong context. Lacking any sort of meeting structure, groups often go through an unfocused and inefficient period developing discussion protocols and group expectations. Many individuals also attempt to promote personal agendas during this initial period of disorganisation, which can bias subsequent group interaction.

2.3 Strategic Research Planning

Developing a long-term research program involves strategic planning. Formal studies of strategic decision-making practices have found that logical and sequential steps are rarely used, sophisticated methods for problem formulation are lacking, and alternatives are not critically examined (Milliken and Vollrath 1991). The four components of strategic decision making or planning (McGrath 1984) were mentioned previously, and include: generating, choosing, negotiating, and executing. The GDM approach described below is a highly structured process that relies heavily on the AHP for its structure (refining and organising), and utilises brainstorming as an idea-generation mechanism. Negotiation (or agreement) is supported within the process but is not required due to the capability of the AHP to calculate an average of disparate judgements. When options (or alternatives) are prioritised with respect to both importance and feasibility, an implementation plan emerges naturally (e.g., select alternatives with high importance and high feasibility). However, we have also supplemented the process with a "strawman document" that acts as an archetypal template to provide initial content for group discussions. Such a document provides the group with a starting point for deliberations, and removes much of the time-consuming, procedural gymnastics that groups experience while trying to develop an operational protocol for discussion.

We illustrate the application of an AHP-based GDM process in a strategic context by formulating a research program for assessing the effects of large-scale fire disturbances (Schmoldt *et al.* 1999). We developed an AHP-based process for workshop settings based on the success of the AHP in similar group settings (Basak and Saaty 1993, Bryson 1996, Choi *et al.* 1994, Dyer and Forman 1992, Madu and Kuei 1995, Peterson *et al.* 1994, Reynolds and Holsten 1994) and its ease of application compared to multi-attribute utility theory (Bard 1992). The GDM process described here is potentially applicable to many types of workshops, meetings, and other temporary (or persistent), formal group tasks.

3. AHP-BASED GROUP DECISION MAKING

During the past decade, there has been a proliferation of workshops associated with planning and decision making in federal agencies. However, the personal experiences of many workshop participants are that such meetings are often unfocused and unproductive, wasting both time and money, and producing results with little substance. Although the AHP has most often been applied in small-group settings, it is also effective in facilitating the conduct of large workshops that include decision making as a component of their objectives (Schmoldt *et al.* 1999).

Workshops will succeed only if (1) the workshop host has clearly stated the objectives (Silsbee and Peterson 1991, 1993), (2) the workshop process is highly structured, and (3) there are specific products resulting from the workshop. As with any discussion group, size matters, because a group with too many participants leaves little opportunity for any single individual to contribute. Introductory information and plenary sessions should be relatively brief and directly relevant to the objectives of the workshop. One or more facilitators, who are willing to assertively guide the workshop process and keep discussion focused, are a key to successful workshop outcomes.

3.1 Workgroup-Focused Deliberations

While a workshop may have many participants, most of the actual work is best conducted in smaller *workgroups*. Each workgroup can be assigned a discrete part of the overall decision problem—for example, in Figure 1, each workgroup was assigned a single "primary topic". Our GDM process is designed to operate in this intimate, participant-friendly environment of small workgroups. In the context of GDM, each participant has more opportunity and greater willingness to contribute (less introverted behaviour and less social loafing), and social inhibitions are less pronounced. Members of each workgroup can also be given considerable freedom to move about and participate in other workgroups as appropriate (for informational purposes only). This encourages wide-ranging contributions by participants (also hindering introverted behaviour) and facilitates between-group interaction (discourages social loafing). Use of disjoint workgroups is particularly effective when primary topics are relatively focused and discrete. However, care must be exercised when making workgroup assignments, because it is possible to unwittingly skew workgroup membership in a negative or political way.

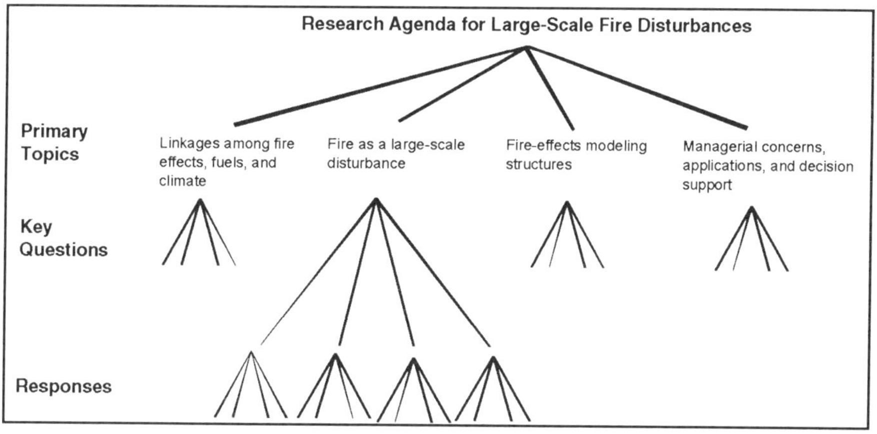

Figure 1. The hierarchical organization of primary topics, key questions, and response to key questions is illustrated. The *response* layer is displayed for only one key question; it would be duplicated for the others. Terminology for each level is generic and designed to accommodate many types of decision problems.

3.2 Strawman Document

It is normally helpful to present workshop participants with a "strawman" document as a framework for discussion and potential revisions (Schmoldt and Peterson 1991). In the case of an inventory and monitoring (I&M) program, the strawman can be a summary of key scientific/managerial questions and responses, sample project statements, or a programmatic plan developed by someone else. The strawman may eventually be completely revised in the course of the workshop, but its presence is extremely helpful in reducing unfocused discussion and as a starting point for initial deliberations.

3.3 Hierarchical Organisation of Topics

In keeping with the overall structure of the AHP, a hierarchical organisation of workgroup discussion topics is used. We can organise this hierarchy using the generic concepts of *primary topics*, *key questions*, and *responses* (Figure 1). These generic terms for hierarchy sub-levels are used because they are intuitively understandable and reflect a problem-solving approach to a technical workshop assignment. Their generic nature also means that the same hierarchical structure and terminology could be used for other technical workshops, or supplanted with more workshop-specific terminology. An initial hierarchy is presented in the strawman document, although workgroups can modify this structure as they develop their own topics. Subsequent levels of each sub-hierarchy contain key questions and responses to key questions.

The hierarchy presented in Figure 1 is not a traditional AHP hierarchy, but rather, more like a taxonomy. In a typical AHP exercise, items at each level are compared pairwise with respect to *each* element in the level above, and priority values are propagated down the hierarchy to alternatives (in this case, responses to key questions) at the lowest level. This produces a *fully-connected* hierarchy, where all items on each level are connected to all items on adjacent levels. For the fire workshop described below, the hierarchy is singly connected, therefore, each response receives only a contribution of importance (or feasibility) from one key question in the preceding level.

Because each workgroup discusses a single primary topic, workgroup sub-hierarchies can eventually be combined to form a global hierarchy for the workshop—each primary topic would be an element on level one of the global hierarchy. Comparisons could then be made among the primary topics according to importance and feasibility. Program managers could perform this step, if importance and feasibility have strategic relevance. However, this level of comparison is beyond the scope of the workgroup context, each of which focuses on a single primary topic.

3.4 GDM Process

With the use of small workgroups, an AHP-based hierarchical structure of discussion topics, and an archetypal template (strawman document) as operational tools, the general process for each workgroup is to: (1) identify key questions in the primary topic area assigned, (2) rank those key questions with respect to importance (and feasibility, where appropriate), (3) articulate responses to each of those key questions, and (4) rank the responses to each key question with respect to importance and with respect to the feasibility of scientific knowledge, models, and data. Because steps 3-4 for *responses* duplicate steps 1-2 for *key questions*, the next two sections refer to them both as "issues" and they are not duplicated here for both types of issues.

3.4.1 Idea generation

One of the most familiar GDM techniques, brainstorming, has been around for a long time. It simply provides for face-to-face discussion between individuals with the intent of generating ideas. In a round-robin fashion, group members offer ideas, which are recorded for later discussion. Ideas that seem to have a nominal amount of group agreement are eventually retained (McGrath 1984). Brainstorming is valuable for making lists of things and generating ideas. However, individuals working alone can generate more ideas than when working in groups, which suggests that group dynamics can have a negative impact on brainstorming (Lamm and Trommsdorff 1973).

Because brainstorming aims to generate lots of ideas, workgroup members offer up issues while someone records them. Brainstorming can use the strawman document as a template for generating ideas or can be done independently of the strawman. In any case, the objective is to generate many issues as quickly as possible. No evaluation of issues is made at this point; rather, judgement is deferred until subsequent discussion. When the production of additional issues begins to dwindle, further enumeration is suspended and discussion commences.

Issues identified by brainstorming can be further refined during discussion. Workgroups can augment each issue to include a clear statement of its meaning and a thorough explanation of its rationale and its position within the primary topic. Recorders then edit these descriptions as necessary and can print out copies for all workgroup members to reference in subsequent deliberations.

3.4.2 Issue ranking

The AHP is used to prioritise and rank the individual issues within each list generated by each workgroup. As described above, this is conducted by all workgroup members (who make pairwise comparisons of the issues), with final scores calculated for the group as a whole. Geometric averaging should be used for these ratio-scale judgements. Individual rankings should generally be compiled privately by each person to avoid the possibility of biases. It is recommended that rankings be developed for both importance and feasibility (or practicality), in cases where these different criteria have different implications for program development or decision making. By having AHP software available at the workshop, all the raw data for pairwise comparisons can be entered and final rankings can be quickly calculated and reported to workshop participants. An I&M example of this brainstorm/discuss/rank procedure appears in Figure 2.

3.4.3 Analysis of priority vectors

Ranking of list items derived from ratio-scale judgements is a critical part of the AHP (Saaty 1980). Within a workgroup, all corresponding judgements are geometrically averaged to produce a single, group judgement for each comparison. This produces a group priority vector. There are two critical questions regarding final priority vectors. One, is there general agreement among workgroup members with respect to their rankings in the priority vectors? Two, are different values within a priority vector really different?

Each workshop attendee can be viewed as a sample from the population of experts on the workshop topic. Because not all experts agree exactly, each priority vector provided by a workgroup member may differ from other workgroup members. One way to be more confident in these uncertain results is to perform statistical tests. Individual judgements can be treated as samples from a population of experts that are independent and identically distributed. The approach that we use is to conservatively apply distribution-free tests that are analogous to tests based on the normal distribution of vector elements (Smith *et al.* 1995). Because distribution-free tests use rank information only (no magnitudes), they may fail to detect significant differences in some cases.

Three common distribution-free tests that are useful in this context are Friedman's two-way ANOVA, the Kruskal-Wallis one-way ANOVA, and Wilcoxon's signed-ranks test. The Friedman two-way ANOVA test analyses the rankings by different workgroup members on each set of items compared. The null hypothesis is that there is no systematic variation in the rankings across items by workgroup members. The Kruskal-Wallis one-way ANOVA test indicates whether there are differences between the elements of a priority vector taking into account all workgroup member judgements. The null hypothesis is that there are no differences. While this test can indicate when differences exist, it does not specify which vector elements are different. To highlight specific differences, the Wilcoxon signed-ranks test is used. A pairwise table of probability values is created which is equivalent to an ANOVA post-hoc test for mean differences. The combination of these three tests allows us to analyse group, and individual, rankings.

Brainstorm

Anadromous fish
Water quality
Phytoplankton
Benthic invertebrates
Aquatic vascular plants
Resident fish
Amphibians
Zooplankton
Ducks and geese

Discuss

1. Need to divide water bodies into lakes/ponds, streams, and reservoirs; different relevant I&M components for each category
2. Water quality is easiest thing to measure
3. Fish populations are difficult to measure
4. Vascular plants should be considered by the terrestrial vegetation workgroup
5. Birds should be considered by the terrestrial fauna workgroup

Rank

Monitoring component	Lakes and ponds		Streams		Reservoirs	
	AHP priority	Ranking	AHP priority	Ranking	AHP priority	Ranking
Anadromous fish	*	—	0.240	2	*	—
Resident fish	0.212	2	0.205	3	0.460	2
Water quality	0.233	1	0.247	1	0.540	1
Amphibians	0.171	3	0.148	5	*	—
Phytoplankton	0.106	6	*	—	*	—
Zooplankton	0.112	5	*	—	*	—
Benthic invertebrates	0.165	4	0.160	4	*	—

* Resource not monitored in this location

Figure 2. An example of the brainstorm/discuss/rank process for monitoring aquatic biota. Information is typically recorded on a flipchart and/or laptop computer during a workshop (adapted from a workshop for the North Cascades National Park Service Complex).

Although statistical analysis of AHP results provides insight into the decision-making process, a detailed analysis may not be needed for all workshops. If statistical analysis is desired, it should be incorporated in the design of the AHP approach, and someone with statistical expertise should participate in workshop planning and compilation of results.

4. SETTING RESEARCH PRIORITIES: AN EXAMPLE

4.1 Background and Workshop Conduct

The role of fire as a disturbance phenomenon in forest, shrubland, and grassland ecosystems of western North America has long been recognised. Nevertheless, there are many difficulties associated with scientific assessment and management of large-scale fire phenomena. This problem was brought sharply into focus in 1988 during and following the large fires in the Yellowstone National Park region. Given the complexity and importance of large-fire phenomena, there is a need to improve our current scientific assessment and management of natural resources in North America with respect to fire disturbance. In April 1996, a group of scientists and resource managers gathered at the Fire-Disturbance Workshop at the University of Washington to discuss these issues. The workshop objectives were to: (1) identify the current state-of-knowledge with respect to fire effects at large spatial scales, (2) develop priorities for scientific assessment of large-scale fire disturbance and its effects, and (3) develop priorities for assisting scientifically-based decision making with respect to fire disturbance in resource management.

Workshop discussion centred around four *primary topics*: (1) linkages among fire effects, fuels, and climate, (2) fire as a large-scale disturbance, (3) fire-effects modelling structures, and (4) managerial concerns, applications, and decision support (Figure 1). Because these topics are relatively independent, small workgroups were used rather than one large plenary session. Each of the 25 workshop attendees was assigned to one of the four workgroups, based on their established expertise. Both scientists and managers were in attendance—in about a 3-to-1 ratio, respectively.

Following a two-hour introduction to the workshop structure/process (including the use of brainstorming, the AHP, the strawman document, and subsequent analyses of priority vectors), workgroups met for one full day and for two hours on the morning of the third day to discuss and synthesise their results. Total time spent in workgroups was 10 hours. After a morning break on the third day, a plenary session was again convened with a member from each workgroup making a summary presentation to the entire group.

A spreadsheet macro was written to generate matrices and perform AHP calculations during the workshop. The recorder needed only to label matrix-row headings and enter each workgroup member's judgements. The software calculated the priority vectors and consistency ratios. Because all judgements are entered into a spreadsheet, it is then possible to modify selected cells (e.g., judgements) and observe how the priorities and consistency change. Statistical analyses of priority vectors were conducted following the workshop.

4.2 Workshop Results

Experts within a workgroup differed significantly in their ratings for 33 of 48 priority vectors, as determined by Friedman tests that failed to detect a systematic pattern. The workgroups dealing with "linkages between fire effects, fuels, and climate" and "fire as a large-scale disturbance" generally had lower internal agreement on rankings than the other two workgroups. We attribute this effect to the uncertainty and difficulty associated with those two topics (science questions), as well as the more applied nature of the latter two topics (modelling and decision support). In particular, this non-agreement strongly corroborates the feeling that our current knowledge about "linkages among fire effects, fuels, and climate" (primary topic #1) is poorly understood and should be an important focus for future research and expanded modelling efforts (Schmoldt *et al.* 1999). Extensive non-agreement also implies that we avoided the groupthink pitfall, wherein group unanimity bolsters the group against outside criticism. The "managerial decision support" group, consisting mostly of managers, experienced the best agreement (of the four groups) in their rankings. This was particularly noticeable in their importance rankings, although feasibility rankings for future research generated less agreement.

Given the strong non-agreement within workgroups, we suggest limiting the number of workgroup-member judgements used to develop programs and priorities (Schmoldt and Peterson 2000). It is not absolutely necessary to rely on everyone's judgement; certain workgroup members' judgements might be discarded owing to their contributions in other ways (e.g., generating discussion or providing valuable insights). Those same insightful individuals may not necessarily provide good judgements or agree with others.

Because the importance and feasibility of issues interact to determine the foci of research programs, we can plot priority values with respect to those two dimensions. In Figure 3, we consider key research questions only for the "managerial concerns and decision support" group. Intuitively, one would prioritise those key questions that have both high importance and high feasibility, that is high, short-term research priority. In this example, one would choose "communication between model builders and users" based on its relatively high score for both importance and feasibility. Of course, this assumes that equal weight is assigned to both dimensions. Arbitrary lines are drawn in Figure 3 based on an obvious separation between the points in both the importance and feasibility dimensions. As in multi-attribute utility theory, different weights and different mathematical models can be used to calculate the final score.

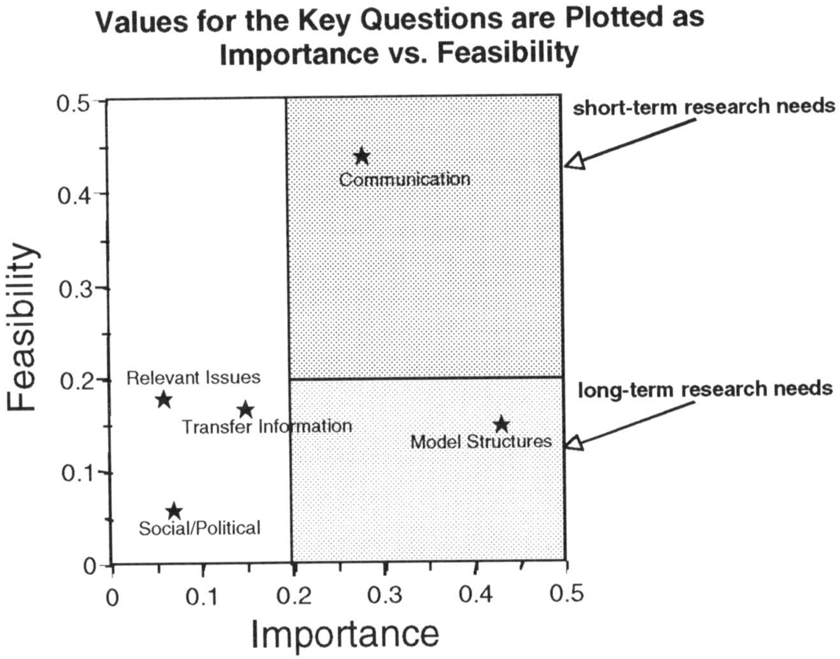

Figure 3. Rating scores for key questions can be plotted according to both importance and feasibility. Those key questions with a high score on both dimensions can be considered good candidates for a research program.

A similar dimensional analysis can be conducted for the responses within each key question. The responses within the highest ranked key question can be examined solely, or global priorities for all responses can be calculated based on the local priorities of key questions and responses.

5. CONCLUSIONS

While our experiences cover several technical workshop efforts, the one described above is the only one to have benefited from a detailed, specific, and rigorous process for GDM. Based on results from all workshops we have facilitated, we can highlight the following ingredients as most critical to workshop success (Peterson and Schmoldt 1999):

- Clearly describe workshop objectives and distribute them and other relevant materials to participants *before* the workshop.
- Limit attendance to no more than 50 people for effective group dynamics; a maximum of six people per workgroup will greatly facilitate decision making. A combination of scientists and resource managers works best, and substantial participation by personnel from the host agency ensures local ownership of workshop output. Resource managers generally are more amenable to using the AHP and less argumentative than scientists.
- Allow movement of individuals between workgroups to promote sharing of expertise and to help develop linkages between related topics.
- Develop a clearly defined product from the workshop output (Davis 1989, Schmoldt and Peterson 2000). This product might be an action/implementation plan or a comprehensive policy statement or a scientific paper. Post-meeting follow-up will ensure that attendees know that something tangible resulted from their hard work, and they will be more likely to participate in future, similar efforts.

A highly structured workshop can elicit a large amount of expert knowledge in a short amount of time. We have found that two days is sufficient to produce the basis of an action plan or similar strategic document. Economic efficiency is an important benefit of this GDM process, because each extra day can cost the host organisation several thousand dollars for salaries, travel, and facilities, in addition to potential frustration for participants. Less structured, and consequently more protracted, meetings produce rapidly diminishing returns for attendees' time. Our experience with using the AHP in group settings (Peterson *et al.* 1994, Schmoldt and Peterson 2000) is that acceptance of the AHP approach quickly follows initial hesitancy and a brief learning period. Implementing AHP decision making interactively in a group setting, for example by projecting a computer display that shows decisions and scores instantly, helps to engage participants and facilitate rapid decisions. Most participants find that this rapid feedback improves their understanding of the decision-making process and speeds up the process by keeping discussions focused.

Some participants even remark that applying AHP interactively in a group setting is fun.

This GDM method contains all the key components of strategic decision making identified by social scientists (McGrath 1984): generating (ideas are produced in brainstorming sessions), choosing (matrices contain value judgements), negotiating (conflict is handled/mitigated by judgement aggregation, but individual judgements are still retained), and executing (several alternatives are given for implementation plan generation, which emerge naturally from the hierarchy and priority vectors). Despite the apparent breadth of this approach, it is relatively straightforward to implement in workshop settings. For smaller, persistent groups (e.g., resource management staffs on a national forest or park), this GDM process may not need to be followed in complete detail, owing to such a group's regularity and familiarity. The important point is that this process offers many advantages—efficiency, comprehensiveness, rigor, and accountability—that the de facto standard (BOGSAT) cannot equal. Both the responsible organisation and its clientele benefit from decision making based on a quantitative and analytical foundation.

REFERENCES

BARD, J.F. 1992. A comparison of the Analytic Hierarchy Process with multiattribute utility theory: A case study. IIE Transactions 24: 111-121.

BASAK, I. AND SAATY, T.L. 1993. Group decision making using the Analytic Hierarchy Process. Mathematical and Computer Modelling 17: 101-109.

BOTTGER, P.C. AND YETTON, P.W. 1987. Improving group performance by training in individual problem solving. Journal of Applied Psychology 72: 651-657.

BRYSON, N. 1996. Group decision-making and the Analytic Hierarchy Process: Exploring the consensus-relevant information content. Computers and Operations Research 23: 27-35.

CALLAWAY, M. AND ESSER, J. 1984. Groupthink: Effects of cohesiveness and problem-solving procedures on group decision making. Social Behavior and Personality 12: 157-164.

CHOI, H., SUH, E. AND SUH, C. 1994. Analytic Hierarchy Process: It can work for group decision support systems. Computers and Industrial Engineering 27: 167-171.

DAVIS, G.E. 1989. Design of a long-term ecological monitoring program for Channel Islands National Park, California. Natural Areas Journal 9: 80-89.

DYER, R.F. AND FORMAN, E.H. 1992. Group decision support with the Analytic Hierarchy Process. Decision Support Systems 8: 99-124.

HALL, J. AND WATSON, W.H. 1970. The effects of normative intervention on group decision-making performance. Human Relations 23: 299-317

HILL, G.W. 1982. Group versus individual performance: are N + 1 heads better than one? Psychology Bulletin 91: 517-539.

JANIS, I.L. 1971. Groupthink. Psychology Today 5: 43-46, 74-76.

JOHNSON, D. AND JOHNSON, F. 1987. Joining together: Group theory and group skills. Prentice-Hall, Englewood Cliffs, NJ.

KAPLAN, M.F. AND MILLER, C.E. 1987. Group decision making and normative versus informational influence: Effects of type of issue and assigned decision rule. Journal of Personality and Social Psychology 53: 306-313.

LAMM, H. AND TROMMSDORFF, G. 1973. Group versus individual performance on tasks requiring ideational proficiency (brainstorming): A review. European Journal of Social Psychology 3: 361-388.

LEANNA, C. 1985. A partial test of Janis' groupthink model: Effects of group cohesiveness and leader behavior on defective decision making. Journal of Management 11: 5-17.

MADU, C.N. AND KUEI, C. 1995. Stability analyses of group decision making. Computers and Industrial Engineering 28: 881-892.

MAIER, N.R.F. 1967. Assets and liabilities in group problem solving: The need for an integrative function. Psychology Review 74: 239-249.

MAIER, N.R.F. AND SOLEM, A.R. 1952. The contribution of a discussion leader to the quality of group thinking: The effective use of minority opinion. Human Relations 5: 277-288.

MCGRATH, J.E. 1984. Groups: interaction and performance. Prentice-Hall, Englewood Cliffs, NJ.

MILLIKEN, F.J. AND VOLLRATH, D.A. 1991. Strategic decision-making tasks and group effectiveness: insights from theory and research on small group performance. Human Relations 44: 1229-1253.

PETERSON, D.L. AND SCHMOLDT, D.L. 1999. Using analytical tools for decision making and program planning in natural resources: Breaking the fear barrier. Pages 256-262 in D. Harmon (ed.), Proceedings of the 10th Conference on Research and Resource Management in Parks and on Public Lands. The George Wright Society, Hancock, MI.

PETERSON, D.L., SCHMOLDT, D.L., EILERS, J.M., FISHER, R.W. AND DOTY, R. 1993. Guidelines for evaluating air pollution impacts on class I wilderness areas in the California. USDA For. Serv. Gen. Tech. Rep. PSW-GTR-136. Pacific Southwest Research Station, Albany, CA.

PETERSON, D.L., SCHMOLDT, D.L. AND SILSBEE, D.G. 1994. A case study of resource management planning with multiple objectives and projects. Environmental Management 18: 729-742.

PETERSON, J., SCHMOLDT, D.L., PETERSON, D.L., EILERS, J.M., FISHER, R.W. AND BACHMAN, R. 1992. Guidelines for evaluating air pollution impacts on class I wilderness areas in the Pacific Northwest. USDA For. Serv. Gen. Tech. Rep. PSW-GTR-299. Pacific Northwest Research Station, Portland, OR.

REYNOLDS, K.M. AND HOLSTEN, E.H. 1994. Relative importance of risk factors for spruce beetle outbreaks. Canadian Journal of Forest Research 24: 2089-2095.

ROGELBERG, S.G., BARNES-FARRELL, J.L. AND LOWE, C.A. 1992. The stepladder technique: an alternative group structuring facilitating effective group decision making. Journal of Applied Psychology 77: 730-737.

SAATY, T.L. 1980. The Analytic Hierarchy Process. McGraw-Hill, New York.

SCHMOLDT, D.L. AND PETERSON, D.L. 1991. Applying knowledge-based methods to the design and implementation of an air quality workshop. Environmental Management 15: 623-634.

SCHMOLDT, D.L. AND PETERSON, D.L. 2000. Analytical group decision making in natural resources: Methodology and application. Forest Science 46: 62-75.

SCHMOLDT, D.L., PETERSON, D.L., KEANE, R.E., LENIHAN, J., MCKENZIE, D.M., WEISE, D.R. AND SANDBERG, D.V. 1999. Assessing the effects of fire disturbance on ecosystems: A scientific agenda for research and management. USDA For. Serv. Gen. Tech. Report PNW-GTR-455. Pacific Northwest Research Station, Portland, OR.

SILSBEE, D.G. AND PETERSON, D.L. 1991. Designing and implementing comprehensive long-term inventory and monitoring programs for National Park System lands. Natural Resources Report NPS/NRUW//NRR-91/04. National Park Service, Denver CO.

SILSBEE, D.G. AND PETERSON, D.L. 1993. Planning for implementation of long-term resource monitoring programs. Environmental Monitoring and Assessment 26: 177-185.

SMITH, R.L., BUSH, R.J. AND SCHMOLDT, D.L. 1995. A hierarchical analysis of bridge decision makers. Wood and Fiber Science 27: 225-238.

STEINER, I.D. 1972. Group process and productivity. Academic Press, New York.

SUNDSTROM, E., MEUSE, K.P.D. AND FUTRELL, D. 1990. Work teams: application and effectiveness. American Psychology 45: 120-133.

VINOKUR, A., BURNSTEIN, R., SECHREST, L. AND WORTMAN, P.M. 1985. Group decision making by experts: Field study of panels evaluating medical technologies. Journal of Personality and Social Psychology 49: 70-84.

WILLIAMS, K., HARKINS, S. AND LATANE, B. 1981. Identificability as a deterrent to social loafing: Two cheering experiments. Journal of Personality and Social Psychology 40: 303-311.

YETTON, P.W. AND BOTTGER, P.C. 1982. Individual versus group problem solving: an empirical test of a best-member strategy. Organizational Behavior and Human Performance 29: 307-321.

Chapter 8

Prioritizing Criteria and Indicators for Sustainable Forest Management: A Case Study on Participatory Decision Making

G. A. Mendoza
Department of Natural Resources and Environmental Sciences, University of Illinois, Urbana, Illinois USA

R. Prabhu
Center for International Forestry Research (CIFOR) Regional Office, Institute of Environmental Studies, University of Zimbabwe, Harare, Zimbabwe

Key words: Sustainable forest management, criteria and indicators, sustainability index, multi-criteria analysis

Abstract: This paper describes an application of the analytic hierarchy process (AHP) in assessing criteria and indicators (C&I) as measures of sustainable forest management. C&I elements are organised in hierarchical manner around the three general concepts, namely: Principles, Criteria, and Indicators. These elements are prioritised based on their perceived relative importance values. These values are calculated using pairwise comparisons of the C&I elements following the principles of the AHP. Pairwise comparisons were obtained from experts representing various disciplines related to forest management. C&I analysis is done at different levels in the hierarchy. To demonstrate the method, a C&I assessment case study involves a forest located in Kalimantan, Indonesia. A generic set of C&I is used as a benchmark. AHP is used to calculate the relative weights of each C&I, prioritise them, and ultimately select a final set of C&I to be used in assessing the sustainability of the forest.

1. INTRODUCTION

Sustainable forest management has become a significant guiding principle in managing the remaining forests worldwide. The overarching

D.L. Schmoldt et al. (eds.),
The Analytic Hierarchy Process in Natural Resource and Environmental Decision Making, 115–129.

objective of attaining forest sustainability was conceived in response to the dual problem of rapidly dwindling global forest resources on one hand, and the increasing pressure to utilise these resources for a variety of uses, on the other. Sustainable management of the tropical forests, for example, by virtue of the forests' strategic importance (e.g. carbon sequestration, habitat to support biological diversity, etc) and developmental significance (e.g. economic returns to support industrial development) has received worldwide attention because of its generally perceived rapid rate of depletion.

While there is general agreement of sustainability as a forest management goal, the practical means to achieve it are still unclear. In fact, debate is still ongoing among forest management scientists on a number of issues such as: definition of sustainable forest management, characteristic features of sustainably managed forests, factors affecting sustainability, and ways to evaluate and monitor forest sustainability.

Notwithstanding these unresolved issues, a number of initiatives have been undertaken, all attempting to realise the goal of sustainable forest management. One of the most significant of these initiatives is the development of criteria and indicators (C&I) for measuring and evaluating forest sustainability. Mainly through the efforts of international organisations such as the Forest Stewardship Council (FSC 1994), International Timber Trade Organisation (ITTO 1992), a number of C&I have been devised and reported in the literature (SGS 1994, SCS 1994). Some of these sets of C&I have in fact been used as bases for certifying whether or not forests are sustainably managed.

Much of the criticism on the development and use of C&I for assessing sustainability centres on the complexity of the forest ecosystem itself. Clearly, many of the dynamic processes, including biophysical, chemical and physiological functions of forest plants and their environment, are too complex and therefore poorly understood. Hence, indices reflecting these processes are difficult to specify and accurately measure. Cognisant of these inherent difficulties, development of C&I was intended to be broad-based; that is, not be narrowly defined. C&I must encompass a wide range of factors operating at different scales and levels of complexity.

2. CRITERIA AND INDICATORS FOR SUSTAINABLE FOREST MANAGEMENT

C&I are essentially tools that can be used to collect and organise information in a manner that is useful in conceptualising, evaluating, implementing and communicating sustainable forest management (Prabhu *et al.* 1996). Following this definition, C&I can be conceived as consisting of four different conceptual elements organised hierarchically as follows: *Principles, Criteria, Indicators*, and *Verifiers*. The definitions, meaning and relationships of each of the conceptual elements are defined by Prabhu *et al.* (1998) as follows:

Principle: A fundamental truth or law as the basis of reasoning or action. Principles in the context of sustainable forest management are seen as providing the primary framework for managing forests in a sustainable fashion. They provide the justification for Criteria, Indicators and Verifiers. Examples of Principles are:

> *For sustainable forest management to take place "ecosystem integrity must be maintained or enhanced", or*
>
> *For sustainable forest management to take place "human well-being must be assured".*

Criterion: A Criterion can be seen as a 'second order' Principle, one that adds meaning and operationality to a principle without itself being a direct measure of performance. Criteria are the intermediate points to which the information provided by indicators can be integrated and where an interpretable assessment crystallises. Principles form the final point of integration. Examples of Criteria when applied under the first Principle given above are:

> *For ecosystem integrity to be maintained or enhanced, "principal functions and processes of the forest ecosystem must also be maintained"; or*
>
> *For ecosystem integrity to be maintained or enhanced, "processes that sustain or enhance genetic variation must be perpetuated".*

Indicator: An indicator is any variable or component of the forest ecosystem or management system used to infer the status of a particular Criterion. Indicators should convey a 'single meaningful message'. This 'single message' is termed information. It represents an aggregate of one or more data elements with certain established relationships. Examples of Indicators when applied to the above Criterion are:

> *To ensure that processes that sustain or enhance genetic variation are perpetuated we can examine the "directional change in allele or genotype frequencies".*

Verifier: Data or information that enhance the specificity or the ease of assessment of an indicator. As the fourth level of specificity, Verifiers provide specific details that would indicate or reflect a desired condition of an Indicator. They add meaning and precision to an Indicator. An example of a Verifier when applied to the above Indicator:

The directional change in allele or genotype frequencies can be determined via periodic measures of the "number of alleles in the population".

Based on the above definitions of the four major conceptual tools of C&I, it is clear that carrying out a forest sustainability assessment should be done following a hierarchical structure. This framework enables the assessment of sustainability at different levels and geographic scales. Prabhu *et al.* (1998) described this C&I hierarchy as shown in Figure 1.

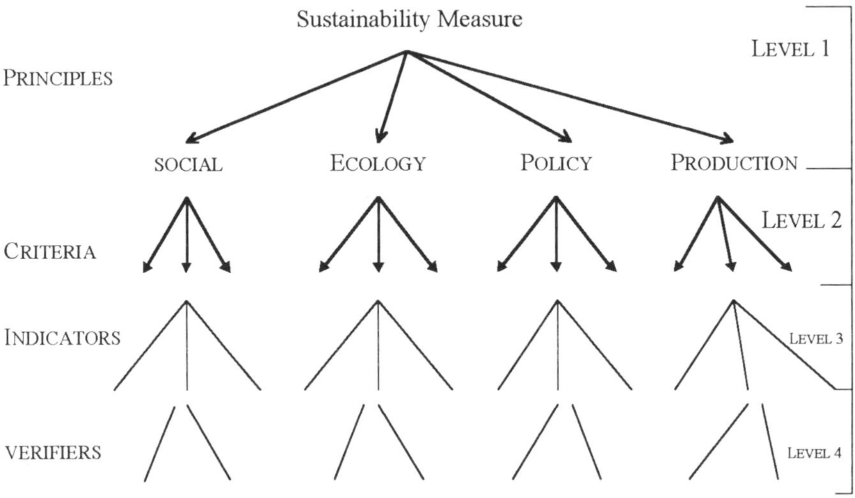

Figure 1. This hierarchical structure of C&I is not complete and is used only to lay out the components of the C&I hierarchy. Hence, the blank boxes are included to denote a set of C&I elements that are too many to include in one figure.

3. THE ANALYTIC HIERARCHY PROCESS

After identifying the four conceptual tools and organising them into a hierarchy as shown in more detail in Figure 2, the next phase of C&I assessment is to evaluate sustainability. As indicated above, the bases of sustainability assessments are the four conceptual tools. However, the process by which these conceptual tools are measured and evaluated, individually and collectively, remains to be established. The section that follows which is derived from Mendoza and Prabhu (1999) briefly describes the procedure and the analytical process.

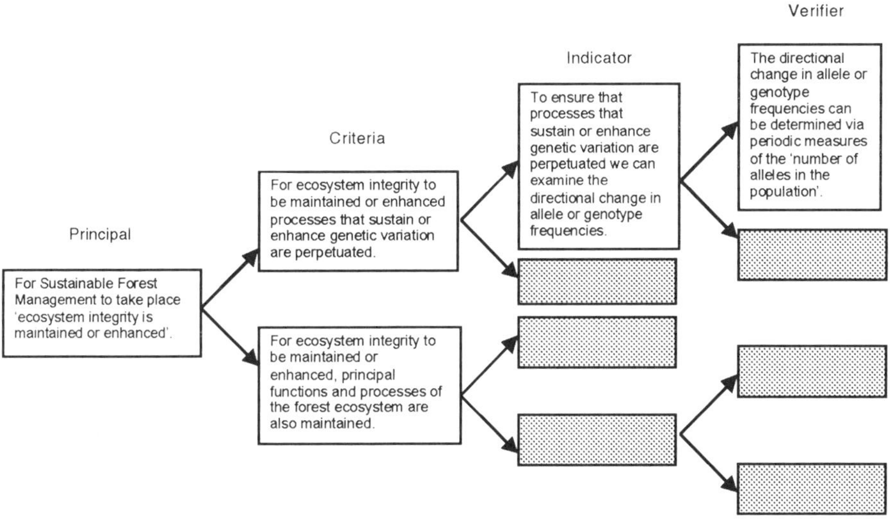

Figure 2. This example of information links in C&I hierarchy is not complete and is used only to lay out the components of the C&I hierarchy. Hence, the blank boxes are included to denote a set of C&I elements that are too many to include in one figure.

The hierarchical structure and multiple criteria attribute of C&I assessment lend itself well to formal analysis using a methodology called the Analytic Hierarchy Process (AHP). AHP can be summarised as a 4-step procedure as follows: Step 1, set up the decision hierarchy by decomposing the problem into a hierarchy of interrelated elements; Step 2, generate input data consisting of comparative judgement (i.e. pairwise comparisons) of decision elements; Step 3, synthesise the judgements and estimate the relative weights; and Step 4, determine the aggregate relative weights of the decision elements to arrive at a set of ratings for the decision alternatives.

Step 1 of AHP involves the construction of a decision problem into a hierarchy of interrelated decisions. At the top of the hierarchy is the goal of the analysis (e.g. selecting the best or most suitable option). The elements at the lower level hierarchies include the attributes such as objectives - perhaps even more refined attributes follows at the next lower level - until the last level which typically contain the options or alternatives.

Step 2 involves the pairwise comparison of the attributes or elements in one level relative to their contribution or significance to the elements of the next higher level. This step constitutes much of the evaluation (quantitative) or assessment (qualitative) of the decision problems and its hierarchy.

However unlike other quantitative decision-making tools, the evaluation and assessment process in Step 2 are easily within the grasp of the decision maker (DM) and the information required of the DM are transparent and are not difficult to provide.

The general principle of comparative judgements in Step 2 is applied in order to construct pairwise comparisons of the relative importance of elements in a level with respect to shared criteria in the level above. Specifically, the input matrix of pairwise comparisons shows the extent to which an element is preferred over the other, or its relative importance and contribution with respect to the element of the level above. In general, the pairwise comparisons are expressed in a scale between 1 (denoting equal importance) to 9 (denoting absolute importance). Intermediate scales between 1 and 9 denote varying degrees of importance from weak to extreme.

The third Step is the synthesis of the judgement matrix described in (3), particularly its square matrix equivalent. With this matrix, Saaty (1996) has shown that solving the primary eigenvector of the matrix will provide an estimate of the relative weights (or eigenvector) of the elements indicating their priority level. That is, the relative weights can be obtained from each one of n of the matrix.

Consider an eigenvector **W** and its elements:

$$\mathbf{W} = (w_1, w_2, \ldots, w_n) \tag{3.1}$$

and the eigenvector equation.

$$\mathbf{AW} = \lambda\mathbf{W} \quad \text{or} \quad (\mathbf{A} - \lambda\mathbf{I})\mathbf{W} = \mathbf{0} \tag{3.2}$$

If there are no errors in measurement (called "inconsistencies" by Saaty,) **A** is considered consistent, and **A** has rank 1; and furthermore, the relative weights of eigenvector **W** could be obtained since $\lambda = n$. In matrix algebra, n and **W** are called the eigenvalue, and the right eigenvector of matrix **A**. AHP recognises that the DM does not know **W**; hence the matrix contains errors and inconsistencies. That is, the DM cannot accurately estimate the pairwise relative weights. However, the estimated weights, **W** can still be obtained using the eigenvector equation in (3.2):

$$\mathbf{AW} = \lambda_{\mathbf{max}}\mathbf{W} \tag{3.3}$$

where A is the observed matrix of pairwise comparisons, λ_{max} is the largest eigenvector of **A** (or sometimes referred to as the principal eigenvalue), and **W** is the right eigenvector which constitutes an estimation of **W**. Saaty (1995) has shown that λ_{max} is always greater than n. The closer the value of λ_{max} to n, the more consistent are the values **A**. Based on this property, Saaty developed the "consistency index" C as:

$$C = (\lambda_{max} - n) / n - 1 \tag{3.4}$$

4. USE OF AHP IN C&I ASSESSMENT OF FOREST SUSTAINABILITY

Assessing forest sustainability is inherently a complex undertaking not only because of its broad scope but also because of the wide range of attributes that bear on its assessment. Operationally, forest sustainability assessments must deal with attributes that are difficult to define and components that may involve both quantitative and qualitative factors. In terms of scope, assessment may cover geographic areas whose boundaries may not be easily identifiable, and socio-economic regions that affect various interest groups or stakeholders each with their own demands and socio-economic needs.

The use of the four major conceptual tools described previously offers a convenient framework with which an organised and systematic assessment of forest sustainability can be carried out. However, with the multiple criteria and indicators involved and the variety of underlying goals and objectives of different interest groups, one might expect that the challenge of arriving at an objective assessment cannot be met using ad hoc procedures. Using simple ad hoc procedures also heightens the risk of generating faulty assessments. Such unfavourable occurrence may be exacerbated by informal decision procedures because they offer little or no "track record" that can help explain the rational or logic employed. This and the lack of transparency of the decision making process can, at best, hinder the adoption of C&I, or at worst, result in failure to gain public acceptance of the results of the C&I assessments.

In many situations, particularly those that can potentially be contentious such as the case in C&I, the ability to communicate and explain the decisions and how they were reached is as important as the decisions themselves. AHP's ability to disaggregate the decision elements and track down the decision making process make it ideally suited for communicating the basis of all decisions.

As described in the previous section, AHP is based on pairwise comparisons of elements such as the C&I conceptual tools organised in a hierarchy shown in Figure 2. Clearly, in each level, it is possible to make pairwise comparison of C&I elements within the hierarchy. Given these pairwise comparisons, relative weights can be estimated based on Equation (3.3). Relative weight is one obvious measure of the importance of each C&I element relative to the node (or parent) in the next level of the hierarchy. That is, within a level, one can estimate a "sustainability index or value" associated to the node using the model below:

$$S = \sum w_i f_i \tag{4.1}$$

where S is a measure of the sustainability value of node i, f_i is the value (score) of lower level C&I element, and w_i is the relative weight of the lower level C&I element I ($0 \le w_I \le 1$). The model above can be repeated progressively to estimate the sustainability values of C&I elements at higher levels of the hierarchy. Note that because the relative weights, w_i are normalised between 0 and 1, the S values are also between 0 and 1 because the 'scores' are assumed to be in percent representing the estimated performance or score for the C&I element. For example, a C&I element (e.g. criterion) based on the expert/s' opinion may be in good condition, in which case, the expert/s will assign a high performance score close to 100 percent.

Based on the discussions above, the use of AHP in assessing forest sustainability using C&I can be described as follows: 1) C&I elements within different levels of the hierarchy are judged based on pairwise comparisons, and 2) These pairwise comparisons are used to estimate the relative importance (or weight) of each C&I element using Eqn. (3.3). These relative weights can be used as a basis for prioritising the list of C&I. Or, the weights can also be used to estimate the performance of a forest management unit by determining the "sustainability index value or score" as shown in Eqn (4.1). These index values can be estimated at different levels of aggregation. That is, at the verifier, indicator, criterion, or principle level.

The ability to measure a sustainability index or score for a C&I element (e.g. a principle or criterion) enables the assessment of forest sustainability according to the specific C&I element. For example, if the objective of the assessment is to examine one particular criterion (e.g., ecosystem integrity), then the estimated 'sustainability index value' for this criterion can be investigated more carefully. This avoids overly generalised assessments on all C&I; rather, sustainability measure is specific to a particular element (e.g. one criterion). Conversely, if the intent of the assessment is to estimate sustainability at a higher level, for example at the principle level, the procedure is also able to generate a composite sustainability measure.

5. A C&I ASSESSMENT CASE STUDY

The forest used as the case study for C&I assessment is located in Kalimantan, Indonesia. The forest management unit (FMU) is about 125,000 hectares and has a 1997 annual allowable cut of about 94,800 cubic meters and 2,200 hectares. All log production was allocated to the company's wood processing mills, mainly the plywood mill.

A C&I assessment team organised for this study consisted of 10 members that include: 1) four employees (i.e., full time staff of the FMU), 2) two villagers coming from two villages under the FMU's community forestry program; 3) one academic lecturer, 4) one government employee working with a government's forest research agency, 5) one social scientist who also works for a foreign assisted development project located within the FMU, and 6) one full time employee of CIFOR who is stationed at the FMU site. All team members are very familiar with the history of the FMU, including its management schemes and harvesting regimes. The four FMU employees have been with the company at different lengths of service ranging from 2 to about 10 years. They are involved in various aspects of FMU activities such as: planning, nursery, social forestry programs, and silviculture. Their roles and responsibilities range from nursery operations to planning and community organisation. The expertise of team members not employed by the FMU are: 1) a general forest management scientist and researcher; 2) a lecturer in forest management currently pursuing graduate degrees in forestry; 3) a community organiser with a legal background; 4) a forester with extensive exposure and familiarity to the area; and 5) two villagers who have resided in the area for at least ten years.

Recognising the discrepancy in expertise, educational background and technical capabilities of the team members, it was necessary to have discussions and detailed presentations of C&I and AHP. This was done before individual opinions and judgements were solicited from the assessment. The assessment process is as follows:

- The Generic C&I developed by CIFOR (Prabhu *et al.* 1998) was translated into the local language.
- Discussions, questions and interactions were all done in the local language.
- The Response forms were prepared and translated in advance.
- Briefing documents briefly explaining C&I in general, and AHP in particular were also prepared and translated in advance.
- General instructions on filling the forms were thoroughly explained.

In order to facilitate the voting process the 10-team members were divided into two subgroups. Group 1 consisted of members whose expertise was related to the Policy and Social Principles; Group 2 consisted of members whose expertise are Ecology and Production Principles.

Before voting began the AHP facilitator explained the following:

- The C&I element (i.e. Principle, Criteria, Indicator) being evaluated.
- The hierarchical relationship between the elements being evaluated.
- The role of AHP.
- The type of input required from the team members

The analysis proceeded as follows:

1. The Criteria level analysis was done first. In this way, the team members were introduced to the analysis at the point where the degree of detail and analysis is of sufficient depth and breadth that is within the grasp and comprehension of all team members.
2. The Indicator level analysis followed the Criteria assessment. At this stage, it is likely that each team member has gained better understanding of the assessment process and the C&I. More importantly, this is the level where the team members feel most comfortable as Indicators are less abstract and more empirical than Principles and Criteria.
3. The assessment at the Principle level was done after the Criteria and Indicator level analyses. It was presumed that by analysing the Principles at this stage, the team members are more cognisant of the C&I and AHP, and would be better prepared to do the broad assessment required at the Principle level. At the Principle level the team was not divided into subgroups.

Note that the analysis was done only on three levels (See Figure 1). The assessment team felt that analysis up to the indicator level was sufficient.

6. RESULTS OF AHP APPLICATION

In this paper, the C&I generated by CIFOR (Prabhu *et al.* 1998) was used as an initial set to begin assessing the sustainability of the forest. This set consisted of six general principles, namely: 1) Policy, Planning and Institutional Framework, 2) Maintenance of ecosystem integrity, 3) Forest Management to maintain and enhance fair access to resources and economic benefit, 4) Local communities and other affected parties, 5) Health, welfare, and rights of forest workers, and 6) Production of goods and services. Under these six principles are twenty-four different criteria, and under the criteria, there were a total of ninety-eight indicators. Pairwise comparisons and relative weights for each principle, criteria, and indicators were analysed and reported in Mendoza and Prabhu (2000). Because of space limitations, only the analysis of the principles and the criteria under one of the principles are presented here to demonstrate the use of AHP in C&I analysis.

6.1 Principle Level Analysis

Analysis, discussion, and voting at the principle level were done in a group setting. Hence, the assessment team of 10 members met as a group to discuss the importance of each principle. While the team met and debated the principles as a group, voting was done individually. Typically, voting was conducted one principle at a time and only after discussion of each principle is completed.

The calculated relative weights of each principle based on the pairwise comparisons given by the assessment team of ten experts are shown in Table 1. These weights reflect the relative importance of each principle as judged by members of the assessment team. From Table 1, it is clear that all principles are important. No principle is rated significantly low enough to warrant elimination from the list. Principles 1,2, and 3 are rated slightly higher than Principles 4,5, and 6. Further discussions of these principles revealed that none of the principles should be dropped in the C&I assessment for the sustainability of the forest management unit.

Table 1. Relative Weights of the Principles

Principle	Relative Weight
1	21
2	18
3	18
4	14
5	14
6	15

6.2 Criteria Level Analysis

As pointed out earlier, analysis at the criteria level was done first. Because of the scope and reasonable clarity of each criterion to team members, it was felt that team members are most comfortable to start the C&I analysis at this level. Team members are generally aware of the different criteria and are able to make or provide the necessary pairwise comparison. The team was subdivided into two sub-groups according to their expertise. Then, each sub-group was assigned a set of principles to evaluate. Sub-group members voted only on those criteria under their respective principles; i.e., Group 1 (6 members) voted on all criteria under Principles 1,3,4, and 5, while Group 2 (4 members) voted on all criteria under Principles 2 and 6.

The criteria under Principle 6 (Production and quality of goods and services) were used to illustrate this level of analysis. Results for other principles are reported in Mendoza and Prabhu (2000). There were 6 criteria under this Principle, namely; Forest management unit is implemented on the basis of legal title on the land, recognised rights or clear lease agreements, 2) Management objectives clearly and precisely described and documented, 3) A comprehensive forest management plan is available, 4) The effective implementation of management plan is effective, 5) An effective monitoring and control system, and 6) Equitable distribution of economic rent.

Table 2 summarises the results of the responses generated from the team members. While some criteria were rated lower than others (e.g., Criteria 1, 2,and 6), careful analysis should be exercised before making decisions on eliminating a specific criterion. In fact, it is recommended that no criterion should be eliminated until Level 3 is completed and the indicators have also been analysed as shown in Mendoza *et al.* (1998).

Table 2. Relative Weights of Criteria under Principle 6.

Criterion	Relative Weights
1	11
2	14
3	25
4	22
5	18
6	11

6.3 Sustainability Index for the Forest Management Unit

In addition to prioritising the list of C&I according to their relative importance based on calculated relative weights, one other use of AHP is to determine the sustainability index for the management unit. This index can be viewed as the measure or degree to which the forest is managed sustainably, or alternatively, it can also be used as a measure of performance in evaluating the company charged with managing the forest management unit.

In the case study, the expert team was asked to provide their expert opinion or best judgement on the condition of the forest or performance of the company relative to the different principles, or criteria included in the set of C&I. For this purpose, the scoring guide used is shown in Table 3.

In this part of the case study, the team did the scoring as a group. That is, each C&I element was discussed and then scored by the group. Further discussion ensued if the group had disagreements about the appropriate score for a given C&I. In the end, only one score was used for each C&I element. In this study, scoring was done at the 'indicator' level. This was done because the team felt it was the level at which there was sufficient clarity for the team to make reasonable scores based on field observations.

Table 3. Guide used for assessing forest management unit

Score	General Description
*	Impossible to score at time of assessment; possibly due to lack of information or unavailability of field sample; to be scored at future date
0	Not an applicable criteria or indicator.
1	Extremely weak performance; strongly unfavourable.
2	Poor performance; unfavourable; may be at the norm for the region, but major improvement needed.
3	Acceptable; at or above the norm for good operations the region.
4	Very favourable performance; well above the norm for the region, but still needing improvement in order to be state of the art.
5	"State of the art" in region; clearly outstanding performance which is way above the norm for the region.

Again, due to space limitations, only the performance scores or sustainability values of the criteria under Principle 2 (i.e. Maintenance of Ecosystem Integrity) is reported in this paper as shown in Table 4. These values were estimated using Eqn (5) where f_i represents the "scores" of each indicator using Table 3, and w_i is the estimated relative weight of the same indicator. Mendoza and Prabhu (2000) contains the details of the scores on all indicators and the cumulative scores for all criteria.

Table 4. Scores or sustainability performance values of forest management unit on criteria under Principle 2 (Maintenance of Ecosystem Integrity)

	Criteria	Sustainability Values	
		Average	AHP-based
1.	The forest management unit has prepared environmental impact assessments.	2.33	2.45
2.	The processes that maintain biodiversity in managed forests are conserved	2.33	1.99
3.	Soil and water processes are maintained	2	2
4.	Chemical contamination of forest resources is eliminated, or at minimum, reduced to minimum level	3.5	**
5.	The forest management unit supports research documenting. The richness/diversity of selected species groups.	3	3

**AHP-based value could not be calculated because only two indicators were present under this criterion.

From Table 4, it could be observed that the use of AHP can yield significantly varied result compared to simple 'averaging' of the indicator scores. Judging from the relative weights of the different indicators, it is clear that some indicators are deemed highly significant when it comes to assessing sustainability of the forest management unit. Without using AHP and the estimated relative weight, it would be impossible to reflect the poor performance of the management unit relative to an important indicator. In this situation, a forest management unit can mask its poor performance on important indicators while doing well on other less important indicators of sustainability. With AHP, all indicators are assigned their relative importance, and hence, the sustainability values shown in Table 4 will reflect these relative weights. For example, in criterion 2, the management unit has a sustainability value of 2.33 if indicator scores are simply averaged relatively good performance based on Table 3. However, if the indicators are weighted via AHP, the cumulative score is 1.99, indicating poor performance.

7. CONCLUSIONS AND RECOMMENDATIONS

This paper has demonstrated that AHP is an effective tool for evaluating and selecting various elements of C&I. The study also showed that AHP offers several desirable characteristics that make it a suitable decision tool for C&I assessments such as: 1) it enhances the participatory approach to decision making where all stakeholders are involved not only as information providers but as decision makers as well, 2) it simultaneously accommodates different criteria, 3) it enables analysis involving mixed data, both qualitative and quantitative, including expert opinions in the absence of 'hard data' and 4) it is transparent to all participants.

From these desirable features and advantages, items 1 and 4 are most significant. Such features make AHP an ideal tool for a 'bottom-up' approach to C&I assessment. Specifically, it offers a democratic and non-threatening approach to C&I assessment at various stages; from the development and generation of initial sets of C&I, to the actual selection and evaluation of final sets of C&I. The simple yet powerful AHP tools provide a decision environment where C&I can be democratically analysed providing ownership of C&I decisions to a wider spectrum of stakeholders, and thereby increasing the chance of acceptance of the decisions emanating from such C&I assessments. Feedback received from the team indicate that despite the wide range of educational backgrounds, they were able to communicate together and articulate the issues surrounding C&I in the context of AHP and its application to the FMU.

Results from implementing AHP at different levels indicate that none of the Principles are rated significantly low enough to be omitted. At the criteria level, some criteria were rated low enough to merit further examination. However, it is recommended that none of the criteria should be eliminated until after the analysis at the indicator level has been completed.

Starting with the criteria analysis affords the team members the opportunity to learn about the C&I hierarchy and at the same time enable them to make judgements at the level that they are comfortable. Finally, the analysis at the principle level in the end provided an opportunity for the team members to synthesise their assessment at the broad context of the six principles.

REFERENCES

FSC (Forestry Stewardship Council). 1994. Forestry Stewardship Principles and Criteria for Natural Forest Management. Oaxaca, Mexico.

ITTO (International Timber Trade Organisation). 1992. Criteria for the Measurement of Sustainable Tropical Forest Management, Yokohama, Japan.

MENDOZA, G. A. AND PRABHU, R. 1999. Integrating multiple criteria decision making and geographic information systems for assessing indicators of biodiversity conservation. Working Paper - GAM 99-1. Department of Natural Resources and Environmental Sciences, University of Illinois

MENDOZA, G.A. AND PRABHU, R. 2000. Development of a methodology for selecting criteria and indicators of sustainable forest management: A case study in participatory assessment. Environmental management. In press.

MENDOZA, G.A., PRABHU, R., SUKADRI, D., PURNOMO, H. AND HARTANTO, H. 1998. Manual: Using multi-criteria analysis as decision tool for applying the cifor generic C&I template. Center for International Forestry Research (CIFOR). CIFOR Special Publication. Bogor, Indonesia.

PRABHU, R. , COLFER, C. AND DUDLEY, R.G. 1998. Guidelines for developing, testing and selecting criteria and indicators for sustainable forest management. CIFOR Special Publication.

PRABHU, R., COLFER, C.,. VENKATESWARLU, P., TAN, L.C., SOEKMADI, R. AND. WOLLENBERG, E. 1996. Testing criteria and indicators for the sustainable management of forests. Phase i. Final report. Center for International Forestry Research (CIFOR). CIFOR Special Publication. Bogor, Indonesia.

SAATY, T. 1995. Decision Making for Leaders: The Analytic Hierarchy Process in a Complex World. RWS Publications.

SAATY, T. 1996. MultiCriteria Decision Making: The Analytic Hierarchy Process. University of Pittsburgh.

SCS (Scientific Certification Systems). 1994. The Forest Conservation Program: Programme Description and Operations Manual. California.

SGS FORESTRY. 1994. Assessors Handbook, Policy Document and Procedures Manual. SGS Forestry, Oxford, UK.

Chapter 9

Integrating the AHP and HERO into the Process of Participatory Natural Resources Planning

Jyrki Kangas and Leena A. Hytönen
Finnish Forest Research Institute, Kannus Research Station, Kannus, Finland

Teppo Loikkanen
Finnish Forest and Park Service, Finland

Key words: Decision support, forest management, planning theory, public participation

Abstract: The analytic hierarchy process (AHP) is finding more and more applications in the participatory processes involved in planning focusing on natural resources and as a tool in environmental decision support. In the forestry context, a heuristic optimisation method called HERO has been used in tactical planning with multiple participants and interests. HERO can make use of pairwise comparisons and the eigenvalue technique in the formulation of the optimisation problem. In this article, the AHP and HERO are discussed from the viewpoints of criteria given for effective participation. A problem of central importance related to the use of the AHP and HERO is that both of them are rather technique-oriented and consequently the method easily affects the process. The combined use of the AHP or HERO and a "softer" planning approach called positional analysis (PA) is suggested as an approach enabling the user to better meet the criteria of an effective participatory planning process. In this hybrid approach, PA provides the overall framework for the participatory process, into which the AHP or HERO is integrated as a decision-support tool. This approach enables the fulfilment of efficient participation from the viewpoints of both technical decision support and the planning process as a whole.

1. INTRODUCTION

Participatory forest planning is defined in this article as a multi-objective forest planning procedure incorporating private individuals' and/or interest

D.L. Schmoldt et al. (eds.),
The Analytic Hierarchy Process in Natural Resource and Environmental Decision Making, 131–147.

groups' opinions and objectives concerning forest management into the planning process. The term participation refers to what may be called direct institutional participation, where the participation process is initiated either by public institutions (i.e. public agencies) or private industries. Other forms of participation include indirect institutional participation through parliamentary or corporate channels (e.g. parliamentary elections or labour union negotiations) and non-institutional direct participation (Paldanius 1993). The parties involved in participatory forest planning have opportunities for promoting their own objectives.

There are several possible purposes for public participation. The main purpose is to transmit individuals' preferences into decision making. A common goal is also to inform people about decision-making touching upon matters of common interest. Thirdly, the purpose can be to rationalise decision making by collecting new ideas, alternatives, and information from people affected by the planning. Fourthly, it could be simply a matter of supporting decision making, with no intention to really give decision-making authority to the participants. This being the case, the aim might be to produce information for the decision maker on people's opinions and on the effects that taking them into account might have on the choices; in this way, the decision maker would be better informed of the consequences of alternative decisions. This kind of information is necessary in situation wherein adaptive conflict management is involved, for instance. No single participatory technique, per se, is sufficient for attaining all the objectives of participatory planning. The "best" participation technique depends on the planning situation and the objectives set for the planning process (e.g. Glass 1979).

Unfortunately, public participation could also have the purpose of manipulation, with participation being arranged merely for PR purposes. In such cases, participation is often intended either to keep people ignorant, but yet happy about how the planning process is proceeding, or only to meet the minimum legal requirements without there being any real opportunities for participation, and at the same time the real value decisions are being taken regardless of the participatory process (Loikkanen *et al.* 1999).

Public participation also functions as a learning process for the affected community concerning interests about the potential benefits of the proposed action, the alternative courses of action, and their respective consequences (Burdge and Robertson 1990). Moreover, public participation has a potential for addressing the need for reconciliation of various conflicting resource uses (Knopp and Caldbeck 1990). It can be taken as a preventive overall planning approach for managing conflicts before the situation gets out of control and negative impacts accumulate for both the stakeholders as well as the community at large.

Several studies and applications have been published on the use of the analytic hierarchy process (AHP) in participatory natural resources planning and environmental decision support. In the forestry context, a heuristic optimisation method called HERO has also been used in participatory and multi-objective planning. In this methodologically-oriented article, some practical applications of the AHP and HERO are first briefly presented. Then the AHP and HERO are discussed from the viewpoints of criteria given for effective participation. These criteria are two-fold. The efficiency of the methods can be studied from the viewpoint of applying them as technical decision-support models used in a participation process. The methods of participatory planning can also be assessed from more theoretical aspects: Is there a correspondence between the properties of the method and the theories relevant to public participation? The latter criteria are mainly process-oriented ones. Concerning them, the crucial points are how well the method supports the participatory process, and how it can be utilised to better meet the process-wise purposes of participation.

The combined use of the AHP or HERO and a "softer" planning approach called positional analysis (PA) is suggested as an approach to meeting the criteria of an effective participatory planning process better than by applying the AHP or HERO alone. In this hybrid approach, PA, as a method justified from the viewpoint of general planning theories, provides the overall framework for the participatory process into which the AHP or HERO is integrated as a decision-support tool.

2. THE AHP IN PARTICIPATORY NATURAL RESOURCES PLANNING

The AHP (Saaty 1980) was applied in participatory natural resources planning of the Ruunaa Nature Conservation Area, comprising a total of 7,330 hectares, in eastern Finland. The area is state-owned and administered by the Finnish Forest and Park Service (FPS). The task set out in the management plan was to divide the conservation area into two sub-areas: a virgin-forest area and a parkland area. In the virgin area, no silvicultural treatments are allowed. In the parkland, soft treatment schedules emulating natural stand dynamics can be applied, and some recreation management is allowed. For example, natural regeneration of small areas can be carried out, and fireplaces as well as hiking routes can be established within the parkland area.

Six alternatives, the border lines between the virgin and parkland areas being placed along ecological edges, were defined for further consideration. The alternative with the greatest proportion of parkland included about half of the area as virgin area and half as parkland. Correspondingly, the alternative at the other end of the spectrum consisted of virgin area only. In each of the alternatives, the most valuable parts with regard to the conservation aspects were included in the virgin area.

The preferences of fourteen interest groups were analysed. The said interest groups represented government officials at the provincial level, local municipality officials, nature conservationists, representatives of local sports associations and hiking as well as hunting associations, research institutes and researchers from the nearest university, and local inhabitants.

Four criteria were considered: priority with respect to considerations of conservation, priority with respect to research activities, priority with respect to recreational activities, and priority with respect to wood production (Figure 1). The choice of criteria was based on a law issued concerning the area. The same decision hierarchy was applied to all participants. Representatives of each interest group compared pairwise the importance of these criteria. When calculating local priorities of the alternatives with respect to conservation, research, and wood production, comparisons performed by experts on corresponding subject areas were used. Only recreational priorities were estimated separately for each interest group on the basis of comparisons made by the representatives of the groups.

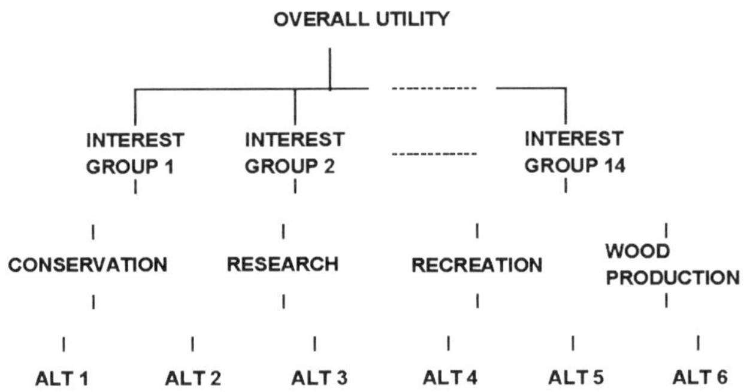

Figure 1. The decision hierarchy of the case study conducted at the Ruunaa Conservation Area (Kangas 1994). All interest groups and all connections are not shown for space reasons.

Most interest groups were of the opinion that conservation-related considerations should form the most important decision criteria in the division of the area into sub-areas. Conservation aspects had the greatest weight from the points of view of eleven interest groups. Only three interest groups found recreation to be the most important use in the area. The starting point in the calculations was to have equal weights for the interest groups, i.e. the weight of each interest group was 1/14. Several weighting schemata were then applied as sensitivity analyses in order to obtain support for decision making.

The decision alternative with the whole area belonging to the virgin part was the most recommendable one from the point of view of ten interest groups. It also proved to have the greatest global priority when equal weights for the interest groups were applied. However, different weighting schemata of interest groups can lead to different rankings. The reasons for the differences in rankings of decision alternatives were the divergence of the opinions of the interest groups concerning the importance of the criteria, and the differences in the recreational priorities of decision alternatives from the points of view of the interest groups.

Pairwise comparisons made by the representatives of the interest groups were fairly consistent. The mean Consistency Ratio (CR) of the comparisons of the criteria was 10.5%, with a maximum of 21.0% and a minimum of 3.4%. The mean CR of the comparisons of the decision alternatives with respect to recreation considerations was 6.8%. The CR of the comparisons made by experts was always less than 10%.

The final choice by the FPS was made in compliance with the alternative in which the whole area belonged to the virgin area but with some recreational services such as hiking routes and fireplaces being included. Consequently, the decision was made in compliance with the opinion of the majority of participants, although no real decision power was allocated to the participants. For the details of the case study, see Kangas (1994).

3. HERO IN PARTICIPATORY TACTICAL FOREST PLANNING

In general, tactical forest planning proceeds in two phases. First, several treatment schedules per each forest stand, i.e. compartment, are produced through computer simulation of the planning period. The purpose in this is to predict stand development under different treatment regimes, and to compute the corresponding removals, costs, incomes, and other relevant variables of interest. Secondly, the optimal combination of treatment schedules is sought through numerical optimisation. With HERO, the optimisation step may be divided into two tasks: estimation of the utility function and maximisation of the utility function. The former task can make use of Saaty's (1977) ratio scale estimation technique.

When applying HERO to participatory planning, the focus is on the values and preferences of the participants in relation to the management objectives for the defined planning area. The applied technique elicits the objectives related to the resource as perceived by the people involved. Then it combines these objectives into an overall utility function. The maximisation phase converts these expressions of utility into an optimal management plan.

The overall utility function of the standard version of HERO takes the following form:

$$U_{tot} = \sum_{j=1}^{n} w_j U_j = \sum_{j=1}^{n} w_j \left(\sum_{i=1}^{m} a_i u_i(q_i) \right) \tag{3.1}$$

where U_{tot} is the total utility; w_j is the weight of participant j; U_j is the utility calculated by the utility function of participant j as estimated using the standard HERO utility model (in parentheses: a_i is the coefficient describing the importance of the objective variable i, $u_i(q_i)$ is the sub-priority function related to the objective variable i, and m is the number of objective variables of the participant in question) getting values between 0 and 1; n is the number of participants.

The participants estimate their sub-priority functions for each relevant objective, and the importance of the objectives. Expertise-based sub-priority functions can be applied as well. The sub-priority function $u_i(q_i)$ of an objective describes the relative utility produced by different amounts of a product or a resource. For details on the HERO techniques, readers are referred to Pukkala and Kangas (1993), or the article by Kangas, Pukkala and Kangas in this book.

An application of the participatory planning approach was carried out with forest inventory data obtained from the Pitkäjärvi state forest managed by FPS. The area covers 1,350 ha in North Karelia, eastern Finland. It was divided into 389 compartments to serve as basic calculation units.

Three distinctive interest groups were called upon to participate in the tactical planning process: the FPS, nature conservationists, and local inhabitants. Each interest group selected objective variables best reflecting their goals. The importance of the objectives was determined on the basis of pairwise comparisons of the variables.

The FPS's utility function was determined to be as follows:

$$U_f = 0.1676u_{1f}(N) + 0.8323u_{2f}(V) \tag{3.2}$$

where N refers to the net income during the 10-year planning period, V refers to the remaining standing volume at the end of the period, and u_{1f} and u_{2f} are their respective sub-priority functions. Both sub-priority functions proved to be clearly concave, thus reflecting decreasing marginal utility.

The conservationists' utility function was estimated through interaction and iteration, and it was finally defined as:

$$U_c = 0.8302u_{1c}(BD) + 0.1698u_{2c}(V) \tag{3.3}$$

where *BD* stands for the biodiversity index at the end of the planning period (i.e., after ten years) and *V* stands for the remaining total volume. The sub-priority of biodiversity was chosen to depend linearly on the biodiversity index, which was computed using the formula presented by Kangas and Pukkala (1996).

The local inhabitants selected four objective variables for their utility function:

$$\begin{aligned} U_p =& 0.4250u_{1p}(H) + 0.1550u_{2p}(V) + \\ & 0.2951u_{3p}(BY) + 0.1248u_{4p}(RS) \end{aligned} \tag{3.4}$$

where *H* is the total harvested volume during 1995 - 2004, *BY* is the estimated mean annual berry yield in the year 2005 (kg/ha), and *RS* represents the mean recreation score of all stands in that year. Harvest removal was taken as an indicator of local employment possibilities, and it was assumed that its utility is directly proportional to the harvested timber volume. The remaining volume was an estimate concerning the cutting possibilities after this 10-year planning period.

The FPS staff responsible of forest management evaluated the weights that should be given to each group when making the final decision. Paired comparisons were applied as an evaluation technique to assess each group's weight in combining the overall utility function. The pairwise comparisons resulted in the following joint utility function:

$$U_{tot} = 0.5389U_f + 0.1669U_c + 0.2942U_p \tag{3.5}$$

where U is the total utility, and U_f, U_c, and U_p utilities computed from the utility function of the FPS, nature conservationists and local people, respectively. Once the weights and the utility functions of the interest groups were agreed upon, the total utility function, amenable to numerical maximisation, was formulated as in (3.6).

In addition to the basic optimisation using the overall utility function, several other calculations were made, e.g. maximising the participants' own utility functions separately. Also, mapping the production possibility boundaries between some of the most interesting objective variables and sensitivity analysis were carried out to provide information for the potential negotiations between the interest groups, and as decision support (Figures 2 and 3). Sensitivity analyses revealed that the optimal plan was not very sensitive to changes in the interest group weights in this case. Accordingly, sensitivity analyses could be used to identify such situations where additional negotiations might prove unnecessary. This is the case when changes in the weights do not affect the optimal forest management plan. These situations were surprisingly common in the case study. The multi-party option of HERO, and the case study, are presented in more detail in Kangas *et al.* (1996).

$$\begin{aligned} U_{tot} = &0.5389 \times \left[0.1676 u_{1f}(N) + 0.8323 u_{2f}(V)\right] + \\ &0.1669 \times \left[0.8302 u_{1c}(BD) + 0.1698 u_{2c}(V)\right] + \\ &0.2942 \times \left[0.4250 u_{1p}(H) + 0.1550 u_{2p}(V) + \right. \\ &\left. 0.2951 u_{3p}(BY) + 0.1248 u_{4p}(R)\right] \end{aligned} \tag{3.6}$$

4. CRITERIA FOR EFFECTIVE PUBLIC PARTICIPATION

The criteria for effective public participation should be set separately for the specific techniques applied and the overall participation process. The process criteria include accessibility (i.e. openness of the process and access to relevant information), fairness (in terms of procedural aspects), comprehensibility (i.e. communication and use of jargon), and empowerment (i.e. perception of the impact each party has on the decision) - as perceived by the participants (e.g. Parenteau 1988, Knopp and Caldbeck 1990, Landre and Knuth 1993). In general, the acceptability of a decision depends on how the process is carried out; it makes a difference how a decision is reached (Lewicki and Litterer 1985). Participants prefer solutions which are understandable and whose grounds are both understandable and acceptable.

Many planning theorists emphasise the importance of the manner in which the planning process is performed (e.g. Healey 1992, Sager 1994). According to them, the definition of the planning problem, for instance, should be made respecting the decision-making context. This means, among other things, that the definition should be problem-oriented, not problem-solving-technique-oriented. It is very important to allow and encourage open discussion about values and norms during the decision-making process. Open discussion presupposes comprehensibility, truth, sincerity, and it should be free of any kind of domination. These thoughts are firmly based on general planning theories, such as Davidoff's (1965) advocacy planning theory. Davidoff considered planning as a process for promoting pluralism in society. Other planning theories appropriate for public participation include that which may be referred to as the transactive planning theory developed by Friedmann (1973) and the communicative planning theory of Sager (1994).

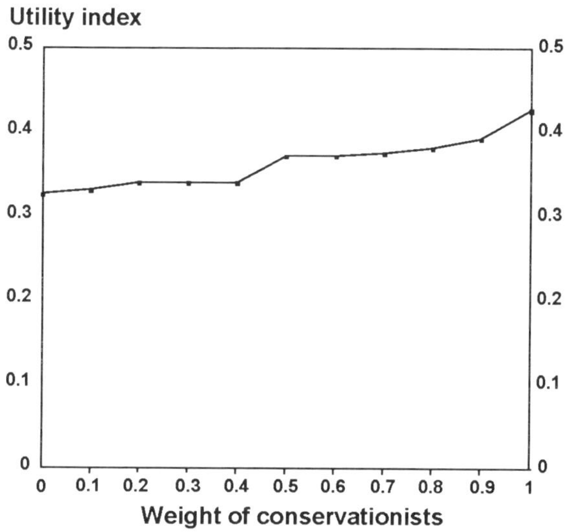

Figure 2. An example of sensitivity analysis using HERO: the effect of changing the weight of conservationists (in the joint utility function) and their resulting utility index (Kangas *et al.* 1996).

According to Knopp and Caldbeck (1990) "participatory democracy exists, when individuals have a known and quantifiable effect (more than zero) on the decision". Furthermore, there should be little room for variation in meaning and manipulation; trade-off decisions among the perceived benefits of the various alternatives should be possible to be made by the participants; and, in order to arrive at a collective decision, individual preferences should be combined in a clear, readily understood manner so that participants know how they have affected the outcome (Knopp & Caldbeck 1990).

Tanz and Howard (1991) suggested taking the following criteria into account when using computer models and technology in forestry decision making: the applications should be easy to understand by non-technical lay-persons; they should represent the forest resource dynamics and be transparent - both objectives and constraints should be easily formulated and modified; they should generate trust in the participants in terms of how they can affect the solution; they should be based on a process that is simple and clear for all parties after a minor introduction, be user-friendly, and be microcomputer-based allowing portability as well as being relatively fast to run; and they should provide outputs in a form that can be interpreted easily.

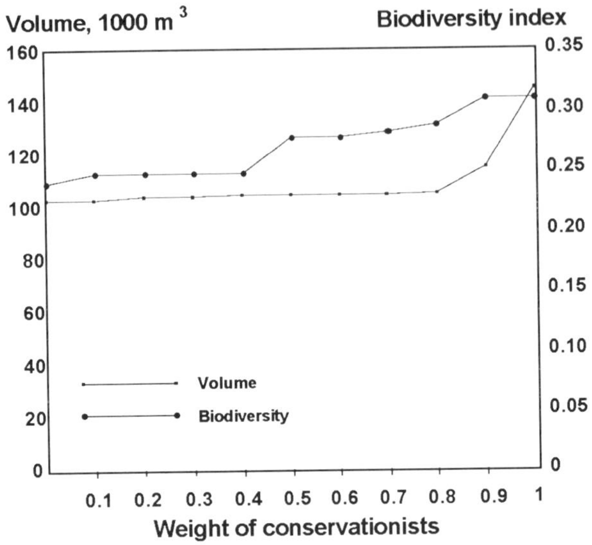

Figure 3. An example of sensitivity analysis using HERO: The effect of changing the weight of conservationists in the joint utility function on the values of objective variables of interest for them (Kangas *et al.* 1996)

There are several other important questions when evaluating methods used in participatory planning from the viewpoint of the whole process. These include the following: Is it possible for individuals to participate in the formulation and delimiting of the decision problem? Is it possible to have two-way exchanges of information at every important step when using the method? Is the way the method manages information suitable for participation by individuals? Is the method (or methods) applied able to deal with all kinds of information?

Two-way exchange of information means that people are informed of and they in turn are given opportunities for stating their opinions concerning planning and decision making. A negotiation situation between the planning organisation and interest groups is a precondition for real participation. The highest level of participation is achieved when people have full decision making authority. It is obvious that for this level to be achieved people or their representatives should be able to participate in planning and decision-making at every important step of the process, from defining the problem to controlling the implementation of the plan.

The approach used should be able to treat both qualitative and quantitative information and values associated with the information. Furthermore, clear and comprehensive analysis should be performed so well, that anyone is able to evaluate the alternatives from his/her standpoint by him/herself, no matter what kind of information s/he has given as input. An appropriate "open" approach would also produce analyses about the benefits and disadvantages of the alternatives from the viewpoints of the different parties. In this way, the process promotes negotiations, communications, and mutual understanding. Following the planning process, there should also be a control and feedback system to help people monitor the implementation of the plan.

As a conclusion based on the case study experiences, the AHP as well as HERO satisfy quite well most of the criteria given by Tanz and Howard (1991) for the models and technology applied in participatory forest planning. However, the final practical benefits of any development are strongly dependent on how well it is integrated with the whole participation process. It is apparent that neither the AHP nor HERO alone can fulfil all the requirements of a perfect participation process. They can mainly serve as the hard core of the process and as decision-support tools. On the other hand, the AHP and HERO do not prevent effective participation as long as they are carefully applied. Matters of importance are what kind of a participatory process the analyses are being produced, how they are utilised, and how the results are presented and interpreted. Because neither the AHP nor HERO can manage all the information involved in a participatory planning process - such as highly detailed and qualitative information - they need a more comprehensive planning approach to be applied within.

A potential problem when applying any decision-support method is that the participants do not understand the results nor how the results have been arrived at. This being the case, the participants must rely upon planners and consultants. Often this confidence is far from perfect and the participants may lose trust in the whole planning process. Therefore, the openness and comprehensibility of the process and the inambigiousness of the interpretations of the results cannot be overemphasised. Calculations and analyses should be used first of all as tools for planners and consultants, working groups, etc. Although the experts can be fascinated in applying the methods, the calculations and analyses will certainly confuse most participants not familiar with them.

Most decision-support tools have been developed on grounds of the needs of practical decision making. Compatibility with planning theories has frequently been somewhat neglected by the developers and appliers of the methods, even though their aim may have been to produce useful means for participatory planning. On the other hand, requirements of the planning practice, too, have similarly been neglected when considering planning theories in the context of public participation. This has lead to situations where theoretically-oriented studies have not met the needs of the practice and practitioners. However, planning theories can serve as guidelines for developing the use of decision-support tools in participatory processes so that not only the criteria for the efficiency of the models and technology, but also the process criteria, can be satisfied.

5. AN APPROACH EMBODYING BOTH AHP OR HERO AND POSITIONAL ANALYSIS

In order to properly and efficiently utilise the results of AHP and HERO analyses, they should be closely linked with the rest of the planning process. All the phases of planning should form an integrated whole. Because the AHP and HERO obviously cannot manage all the tasks of a participation process and information contained in the process, other tasks and information should be able to be dealt with using other methods and techniques. Next, positional analysis (PA), developed by Söderbaum (1987, 1994), is introduced as a comprehensive planning approach focusing on the whole participation process consisting of different analyses, tasks, information, etc. This approach includes guidelines for natural resources management planning, and it can be applied as a blueprint principle for processes utilising numerical decision-support tools as well. It covers all the phases of a planning process, starting from the full description of the situation, including even its historical background, and ending at the final decision. Strong emphasis is given to diverse and deep analyses of activities and interests.

PA aims to be open-minded in its relation to all actors and interested parties, and versatile in its analytical properties. The main purpose of PA is to shed light on the decision situation. No consensus regarding valuation rules, and no certain decision support techniques are assumed or required. In the planning process, participants can refer to valuation rules or standpoints that are relevant for them, and point out conditional conclusions. Depending on the kind of decision situation and the social and institutional context, simplified versions of PA may be considered.

The analysis of interests and conditional conclusions, as included in PA, can be modified as follows to be applied as complementary decision-support tools in natural resources planning. In the analysis of interests, various activities and their probable directions are analysed. The main phases of this analysis are as follows: identification of the activities (of individuals or organisations) that will be influenced by forest management; assessing the most desirable directions of future development for each activity, associated with each interest and related goals; and summarising the effects of implementing each decision alternative on each activity, from the viewpoint of each individual goal and their combinations. (Söderbaum 1987).

Conditional conclusions, which may be based on the analysis of interests, are used to articulate the ethical and ideological aspects of the decision situation. The said conclusions are constructed as follows: if the interests *a*, *c*, and *d*, are important, a certain alternative—say, alternative *A*—should be chosen. For instance, "if you prefer moose hunting over forestry you would choose plan *C* rather than plan *A*." Each decision alternative is deeply analysed: which activities and goals, and their combinations, it supports and which it does not. Participants will face the complexity of the decision situation, and, for example, see how some participants may have conflicting goals and preferences. (Söderbaum 1987).

The two approaches (with the AHP or HERO as a decision-support method, and PA as a comprehensive planning approach) can be combined for practical planning processes so that the ability of PA to shed light on the decision making situation is exploited, and the process-oriented problems of the AHP and HERO are avoided, but their analytic efficiency is utilised. The main phases of a procedure appropriate for public participation in natural resources and environmental decision making could be as follows:

a) Description of the planning situation. Preliminary identification of the relevant actors, interests, interested parties and institutions.
b) Detailed identification of the planning problem. Starting an open participation process with traditional participation means and information gathering channels. Organising the first open meeting. Agreement on the need for the planning process with the public. Reproduction of problem images as stated by different actors and interested parties. Agreeing upon the rules to be followed if no compromise could be gained in the process.
c) General formulation of the problem by means of analysis of interest as applied in PA. Explaining how the decision making process is intended to be carried out in the preliminary stage, and gaining commitment for the approach; modifying the approach if necessary. Forming a planning group. The planning group might include a professional planner, representatives of interested parties, and other individuals. The tasks of the group include working as a link between all the interests and the organisation responsible of planning, taking part in the planning work on a voluntary basis, and controlling the process.
d) Each interest within the planning group creates its own decision hierarchy, and corresponding AHP and/or HERO models, together with

planners. The planner would help to analyse how the different objectives can be integrated or are in conflict with another. The planner together with the members or representatives of the interest parties can form an optimal solution from their point of view. Planning calculations are performed for each interest. As background information on the planning problem, calculations on the area's production possibilities as well as conventional cost-benefit analyses are presented to the participants. All the other information gained through the participatory process so far is analysed, too, especially that of qualitative nature. If found appropriate, the AHP and/or HERO models are also derived representing that information mass.

e) The planning group tries to negotiate a solution. The planner's duty is to present possible compromise solutions and conduct the negotiations. Planning calculations and their results are interpreted, justified, and applied as background information in the negotiation process. New calculations, if necessary, are carried out interactively. AHP and/or HERO calculations are made using their multi-party options with differing weights of the participants so that participants can see the effects of different weighting schemes. Assessments are made on how well each interest's concerns are addressed in alternative solutions, and holistic evaluations and conditional conclusions are carried out according to principles of PA and utilising results of AHP and/or HERO calculations. Especially those activities and goals, and their combinations, are carefully considered, which could not be included in the AHP or HERO calculations.
f) Presenting the results of the working group in an open meeting and in different participation channels (such as newspapers, internet, open houses,...). Gaining feedback from the public. Also, alternative solutions with probable consequences might be presented to the public, especially if no initial consensus has been gained in (e). If a general agreement is achieved, proceed to the next phase. Otherwise, return to phase (e).
g) The planning group agrees on the follow-up procedure. The planner writes a report including conclusions about the standpoints of every interest party. The plan is presented widely for the public.
h) Control of the actual implementation of the chosen plan, as agreed upon in (g). Assessing the need for continuous planning procedures according to principles of adaptive planning. Assessing the need for new planning processes.

In phases (d) and (e) it is important to assess the priorities of the alternative solutions by means of comprehensible factors such as net income, scenic beauty, volume of the growing stock at the end of the period; not only by means of utility indices or priority measures. Visualisation and computer graphics have proved to be useful tools in making the calculations and alternative plans more readily communicated.

An important, but, at the same time, a very difficult task is to analyse all the information gathered through different participation channels. Typically, huge amounts of feedback, opinions, preferences, claims, expertise, local knowledge, etc., in various forms (quantitative/qualitative, general/detailed, spatial/not located...) are obtained during the process. In order to utilise this information mass in the process, it should somehow be organised and systemised. Developing appropriate methods for this task is a challenging task for research on participatory natural resource planning. GIS technology would certainly be useful in this.

In phase (e), the negotiations can be run either via integrative or competitive strategies. However, pursuing collaboration is recommended; e.g. because of the more or less continuous future relationship between the parties and the participants (see Lewicki and Litterer 1985). Interactivity is an important feature not only with the negotiations, and with the use of decision-support methods for facilitating them, but also throughout the planning process. It should be possible to return from any phase to any previous phase.

In principle, the process should result in an acceptable compromise solution. In that case only, the ideal of allocating decision power for the participatory process can be deemed to have completely succeeded. In practical planning, this is seldom the case. If no consensus is gained, the process should continue according to the principles agreed upon in phase (b). If no consensus is gained even then, it is up to the institution to decide what to do.

6. FINAL REMARKS

It can be concluded that decision-support methods such as the AHP and HERO provide valuable analyses and information for the participatory planning process, especially in phases (d) and (e) as presented above. The AHP as well as HERO satisfy well the criteria for the decision-support models applied in participatory forest planning. However, in order to meet the process-oriented criteria for effective participation, much more than mere analytical tools and numerical calculations are needed for. The decision-support tools should be organically integrated into an overall planning and participation framework in order to enable the complete exploitation of the benefits of the tools.

In this article, integrating the AHP and HERO with positional analysis is suggested as one alternative to utilising decision-support tools in a comprehensive process of participatory natural resources planning. This way, criteria of effective participation can be satisfied from the viewpoints of both technical decision aid and the whole planning process. In addition to PA, there are other process-oriented, theoretically justified approaches, which could provide a framework for the use of specific decision-support techniques in participatory planning. For example, utilising the AHP and HERO in combination with what is referred to as the Q methodology (e.g. Barry and Proops 1998) would be worth studying further.

All in all, hybrid methods utilising both numerical decision-support tools and means of more general participation approaches seem to be promising when analysing complex participatory natural resources planning processes.

REFERENCES

BARRY, J. AND PROOKS, J. 1999. Seeking sustainability discourses with Q methodology. Ecological Economics 28:337-345.

BURDGE, R. J. AND ROBERTSON, A. 1990. Social impact assessment and the public involvement process. Environmental Impact Assessment Review 10:81.

DAVIDOFF, P. 1965. Advocacy and Pluralism in Planning. Journal of the American Institute of Planners 41:31-338

FRIEDMANN,. J. 1973. Retracking America: a Theory of Transactive Planning. Ancor Press, Doubleday - New York. 289 p.

GLASS, J.J. 1979. Citizen Participation in Planning: the Relationship Between Objectives and Techniques. Journal of the American Planning Association 45:180-189.

HEALEY, P. 1992. Planning through debate. The communicative turn in planning theory. Town Planning Review 63:143-162.

KANGAS, J. 1994. An approach to public participation in strategic forest management planning. Forest Ecology and Management 70:75-88.

KANGAS, J., LOIKKANEN, T., PUKKALA, T. AND PYKÄLÄINEN, J. 1996. A participatory approach to tactical forest planning. Acta Forestalia Fennica 251. 24 s.

KANGAS, J. AND PUKKALA, T. 1996. Operationalization of biological diversity as a decision objective in tactical forest planning. Canadian Journal of Forest Research 26:103-111.

KNOPP, T. B. AND CALDBECK, E. S. 1990. The role of participatory democracy in forest management. Journal of Forestry 88(5):13-18.

LANDRE, B. K. AND KNUTH, B. A. 1993. The role of agency goals and local context in Great Lakes water resources public involvement programs. Environmental Management 17:191-198.

LEWICKI, R. J. AND LITTERER, J. A. 1985. Negotiation. Irwin.Homewood, Illinois. 368 p.

LOIKKANEN, T., SIMOJOKI, T. AND WALLENIUS, P. 1999. A participatory approach for natural resource managers - A guide book. Metsätalouden julkaisuja, Finnish Forest and Park Service. Kuopio. 91 p.

PALDANIUS, J. 1993. Citizen participation in energy policy. Prospects for development of institutional and non-institutional participation in Finland. National Consumer Research Centre. Eripainoksia 5. 22 p.

PARANTEAU, R. 1988. Public participation in environmental decision-making. Ottawa. Federal Environmental Office. 71 p.

PUKKALA, T. AND KANGAS, J. 1993. A heuristic optimization method for forest planning and decision-making. Scandinavian Journal of Forest Research 8:560-570.

SAATY, T. L. 1977. A scaling method for priorities in hierarchical structures. Journal of Mathematical Psychology 15:234-281.

SAATY, T. L. 1980. The Analytic Hierarchy Process. Planning, priority setting, resource allocation. McGraw-Hill. New York. 283 p.

SAGER, T. 1994. Communicative Planning Theory. Avebury. 288 s.

SÖDERBAUM, P. 1987. Environmental management: A non-traditional approach. Journal of Economic Issues 21:139-165.

SÖDERBAUM, P. 1994. The political economics of sustainability. Positional Analysis as an alternative to cost-benefit analysis. International symposium 'Models of Sustainable Development', Paris, France, March 1994. 12 p.

TANZ, J. S. AND HOWARD, A. F. 1991. Meaningful public participation in the planning and management of publicly owned forests. Forestry Chronicle 67:125-130.

Chapter 10

Environmental Cognition: Contributions from the Analytic Hierarchy Process Toward Construction of Cognitive Maps

Reza Banai

Graduate Program in City and Regional Planning, University of Memphis, Memphis TN USA

Key words: Environmental cognition, cognitive maps, wayfinding, imageability, dominance

Abstract: Despite its vast multidisciplinary applications, the analytic hierarchy process (AHP) has received little attention in environmental cognition literature. AHP properties that relate conceptually and methodologically to developments in cognitive mapping and modelling research are highlighted. The method of paired comparisons, which is at the core of AHP procedure, scale, and calculus of consistency, is used with a cognitive mapping application. A taxonomic concept of imageability provides a survey protocol for classification of the elements of the environmental structure. The connective structure of paired relations among the elements as well as their relative dominance as perceived by subjects is gauged (interpreted) by the consistency index in the construction of a cognitive map. This approach captures the qualitative, topological properties of spatial structure while it accounts for observer variation as well as degree of consensus in the image(s) of structure. The chapter concludes with promising future areas of research and development of the AHP as a paradigm for environmental cognition.

1. INTRODUCTION

Emblematic of an applied mathematician, Saaty's introduction to the analytic hierarchy process (AHP) would normally be accompanied by examples of real world applications. One of the early, instructive application examples (reproduced in Saaty 1996, P. 38), involved a simple experiment in which two young children (ages 5 and 7) and one adult were the subjects, who were to stand by a light source and look at the brightness of four

D.L. Schmoldt et al. (eds.),
The Analytic Hierarchy Process in Natural Resource and Environmental Decision Making, 149–165.

identical chairs (denoted by A_9, B_{15}, C_{21}, D_{28}) arranged in a line at various distances (9, 15, 21, 28 yards) from the light source, and to compare their relative brightness in pairs. The judgements of relative brightness are given by the following matrices:

$$\begin{array}{c} \\ A_9 \\ B_{15} \\ C_{21} \\ D_{28} \end{array} \begin{array}{c} \begin{array}{cccc} A_9 & B_{15} & C_{21} & D_{28} \end{array} \\ \begin{bmatrix} 1 & 5 & 6 & 7 \\ 1/5 & 1 & 4 & 6 \\ 1/6 & 1/4 & 1 & 4 \\ 1/7 & 1/6 & 1/4 & 1 \end{bmatrix} \end{array} \qquad \begin{array}{c} \\ A_9 \\ B_{15} \\ C_{21} \\ D_{28} \end{array} \begin{array}{c} \begin{array}{cccc} A_9 & B_{15} & C_{21} & D_{28} \end{array} \\ \begin{bmatrix} 1 & 4 & 6 & 7 \\ 1/4 & 1 & 3 & 4 \\ 1/6 & 1/3 & 1 & 2 \\ 1/7 & 1/4 & 1/2 & 1 \end{bmatrix} \end{array}$$

Relative brightness matrix (first trial, children) — Relative brightness matrix (second trial, adult)

A scale of absolute numbers (1-9), and their reciprocals, is used to quantify the judgements. The eigenvectors of the above matrices—the robust method of estimation of the relative weights of the elements in the AHP—are given below as the synthesis of the judgements of relative brightness of the chairs, and juxtaposed with the results calculated from the inverse square law of optics (illumination intensity decays with the square of distance) in Table 1.

Table 1. Relative brightness eigenvector estimates vs. law of optics

Chairs	Relative Brightness [a] Eigenvector (first matrix)	Relative Brightness [b] Eigenvector (second matrix)	Inverse Square Law (approx.) [c] $I = d^{-2}$
A_9	0.61	0.62	0.61
B_{15}	0.24	0.22	0.22
C_{21}	0.10	0.10	0.11
D_{28}	0.05	0.06	0.06

[a] $\lambda_{max} = 4.39$; CI = 0.13; CR = 0.14
[b] $\lambda_{max} = 4.1$; CI = 0.03; CR = 0.03
[c] See Saaty (1996) for details

It turns out that the judgements have replicated a natural law: Could they do the same in other areas of "thought and perception" Saaty (1996) commented. Saaty seems to have anticipated in advance the wide-ranging disciplinary applications of the AHP in the 1980s and 1990s since the inception of the method in the 1970s. Parenthetically, the example above is interesting in the light of the recent, larger philosophical debates, particularly with the critiques of positivism (e.g., Bhasker 1989) that have questioned the appropriateness of the objective scientific methods of natural/physical systems in social/human systems, which are characterised as "open" rather than "closed" systems and in which human judgement and subjectivity reign supreme. As Saaty (1990, p. 98) has emphasised, "we always interpret other stimuli with our senses—such as how bright light is to the eye or how soft velvet is to the touch. The basic problem is to create a scientific framework for interpreting data."

Noteworthy in the above example is the idea that a scale (of relative brightness) is derived directly with subjects' sensation and interpretation of the (visual) stimuli as (measurement) data rather than a scale that is determined independently of the subjects. But even when the measurement scale is determined objectively, it has no significance in itself until it is perceived, interpreted, understood, experienced, and learned subjectivity. Among the vast, multidisciplinary, and prolific literature of the AHP are the relatively recent contributions in conjunction with geographical information systems in which the appeal of the AHP is with the interpretation of both spatial and non-spatial (attribute) data in the face of uncertainty, diversity, and multiplicity (e.g., Malczewski 1996, Banai 1998). The AHP has not received attention in the cognitive mapping literature despite certain key conceptual and methodological areas that it has in common with those, addressed from either the geographical or the psychological perspective. The plausibility of the AHP is with a psychometric (ratio) scale, which corresponds directly to the interpretative, experiential, and learning tasks performed by subjects in the construction of cognitive maps.

2. TOWARD CONSTRUCTION OF COGNITIVE MAPS: CONTRIBUTIONS FROM THE ANALYTIC HIERARCHY PROCESS

A cognitive map is a representation of the information that a subject receives from external stimuli and stores in long term memory (Garling *et al.* 1998, Golledge 1992; for a review of definitions and concepts, see Kitchin 1994). A cognitive map is constructed by the availability of information, levels of experience, familiarity, learning (knowledge acquisition), repeated experience, and exposure to the sources of information (see Garling and Golledge 1989). Psychological concepts of learning and information processing, geographical concepts of mapping and representation, economic concepts of decision making and choice, computational concepts of information processing and representation, and architectural concepts of use and control, form and function provide useful multidisciplinary insights into the construction of cognitive maps fruitfully at intersections of multiple disciplines. A comprehensive survey of cognitive mapping and modelling research is not intended here (q.v., Garling and Golledge 1993). Our aim is to identify the AHP properties that intersect with conceptual and methodological issues encountered in the cognitive mapping and modelling literature, and thus provide a stimulus for further research and development of a plausible, alternative paradigm for environmental cognition. A brief discussion of the paradigm issues is presented next, followed by an application of the AHP in the construction of a cognitive map.

2.1 Information Processing and Decision Making

Thought of as a "device" in which information is acquired, learned, stored, processed, maintained, updated, and applied to perform a purposeful activity, a cognitive map ironically connotes a computer-like rather than (cartographic) map-like representation. The difficulty exhibited by subjects with distortion or loss of information in a sketch map as a "fragmented" (re)presentation of mental image of a place is suggested by the irony (see Lynch 1960). This conception of a cognitive map does not negate representation of the mental images of space by subjects in cartographic terms. As Golledge (1993, p. 31) indicates, people use the same cartographic notations (point, line, area) to connote graphically "what they know of a place". What is more, people exhibit a greater facility with the knowledge of the topological organisation, or relations of the environmental features rather than with the knowledge of direction and distance (see also Montello 1991, Montello and Frank 1996). The question of how people come to know of a place, in part or as a whole, or find, decide a path to a destination in their routine or non-routine activities, and learn about spatial configuration (or layout) of a place and perceive the frequency and quality of topological, environmental features is arguably more complicated.

2.2 Route Choice and Wayfinding as Multicriteria Decision Making

People learn about place or gain knowledge of spatial configuration in part through experience of paths, and thus wayfinding has been studied from a variety of psychological and geographical perspectives. Studies of children and adults indicate that route knowledge precedes knowledge of environmental configuration (Blades 1991). How do people arrive at a decision about the choice of a path in a network? Optimisation methods state the problem as one of determining the best solution (e.g., shortest path or least cost) subject to constraints (e.g., time or cost). As Golledge (1993) points out, however, we lack the evidence that people's decision about the choice of a route is governed by the same criteria and constraints as those modelled by the linear programming algorithms of operations research. "The potential optimising functions," as Gluck (1991, p. 125) notes, "are not restricted to minimum distance or even minimal effort" for various "logistical and affective reasons." A longer path may be "optimal" if convenience and security are considered jointly with travel time or cost—multicriteria, rather than a single criterion optimisation problem. Furthermore, econometric studies indicate that people consider tradeoffs among the factors that influence travel choice. Included are a combination of tangible (e.g., time, cost) and intangible (e.g., convenience, security) factors. The qualitative factors are generally treated categorically, however, owing to the restriction imposed by econometric (e.g. hedonic) methods with interval- or ratio-scaled factors. The multidimensional, dynamic, adaptive properties, although acknowledged, are not generally accounted for in models of human wayfinding (Gluck 1991). The AHP provides a versatile ratio scale to measure tradeoffs among a diverse set of factors some of which e.g., comfort and convenience, safety and security do not have known measurement scales. But even for factors with known measurement scales e.g., time and cost, the AHP provides a facility to derive utility (or disutility), or to measure perception of (relative) importance to individuals (Saaty 1995).

Consider the route selection problem once again from the perspective of bounded rationality (Simon 1981, 1982, 1983), which is regarded plausibly in the cognitive mapping and modelling literature (Garling *et al.* 1998). People have limited knowledge of the entire travel network. Paths selected are a limited subset of the total "feasible" choice set. People's use of a path is a function of their familiarity with experience over time (dynamic rather than static optimisation). The perception of favourable conditions encourages repeated or routine use of a path. Conversely, people are "tolled off" with the experience or perception of unfavourable conditions (e.g., congestion, accident) and seek alternative paths. Furthermore, the perception or valuation of the relative importance of the multiple travel criteria or factors varies among people as distinguished, for example by class, age, and gender.

An AHP-aided cognitive model of a route selection problem would involve a limited number of criteria and a correspondingly limited set of alternatives (paths)—in contradistinction to combinatorial techniques—that are compared in pairs and their relative merit is weighted by the relative importance of the criteria and summed across the alternatives to determine the likelihood of each chosen path. An application of this process for mode choice problem of interurban travel demand (modal split) produced the following estimates of four travel modes (Figure 1, Table 2. For details, see Banai-Kashani 1984; for urban travel, see Banai-Kashani 1990):

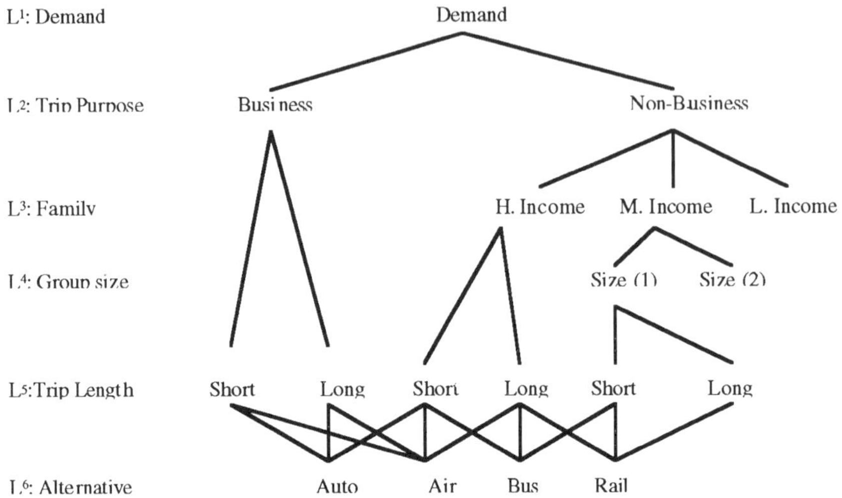

Figure 1. A hierarchy for inter-city travel demand (modal split).

Table 2. Observed and estimated modal split.

	Observed (NCT, 1969) (1968 mill. pass. miles)	Observed Normalised	Estimated Eigenvector
Auto	4226	0.62	0.66
Air	1391	0.20	0.17
Bus	432	0.06	0.06
Rail	767	0.11	0.10
Total	6815		

Among the now burgeoning applications of the AHP are those in environmental resource suitability analysis and spatial decision making, particularly with geographic information systems. Recently, the AHP has been increasingly used as an alternative to the Boolean classification systems particularly in situations involving evaluation of multiple criteria with limited information, uncertainty, and factor diversity. The AHP is now among the recent introduction of methods that similarly employ fuzzy logic and approximate reasoning (e.g., Frank 1991) in geographical information systems. While akin to the concepts of fuzzy sets (Zadeh 1990), the AHP offers an operational alternative to derivation of membership functional values of a diverse and situational variable set of factors without the cumbersome calibration or parametric restrictions imposed by specification of a function (see Saaty 1978, Banai 1993). Furthermore, the fuzziness that arises in decision environments is dealt with as a fundamental, hierarchical structural property of the AHP.

2.3 Hierarchical Structure

Because the concept of a hierarchy is an inclusive, systemic one it is marked by wide-ranging application, from familiar, decision making in organisations to modelling built-environment, individual and collective choice, from computational methods of information processing, to physical or natural systems. As Simon (1981) has noted, complex systems, both human-made and natural, have in large part exhibited hierarchical structures. The concept of hierarchy is also used in models of spatial cognition.

There is evidence of a hierarchical organisation in place recognition (McNamara 1986, 1991, McNamara *et al.* 1989). In a study by Hirtle and Jonides (1985) subjects exhibited an awareness of the specific places throughout a city, which were hierarchically distinguished in relation to a set of dominant landmarks. The study by Peruch *et al.* (1989) also indicates the public awareness and use of the hierarchical organisation of the functional transportation network (freeway-highway-arterial-collector street-local street system). McDonald and Pellegrino (1993) conclude that spatial memory has a hierarchical structure. Golledge and Stimson (1997) note that people are knowledgeable of the hierarchical organisation of spatial configurations if they exist, or, interestingly, mentally construct hierarchies even if the spatial configuration (or layout) itself is not a hierarchical one. In either case, people naturally deal with the complexity of the sensory information and acquire configurational knowledge hierarchically.

The idea of factorability or decomposability of a system, however dense the network of connections, suggests the behavioural plausibility of the AHP's independence axiom akin to the principle of bounded rationality (see also Miller 1956, Saaty 1986). As Simon (1983, pp. 17-18) remarks, "At the moment you are buying a car, you are probably not also simultaneously choosing next week's dinner menu, or even deciding how to invest income you plan to save." And even in collective action, as in organisations, the limits of information and cognition favour strategies that are incremental and partial, rather than simultaneous and total (see also Lindblom 1959). Concomitantly, incrementalism is a behavioural choice in environmental knowledge acquisition as Gluck (1991, p. 118) remarks. Initially, knowledge of "particular locations" or "points" which are not yet perceived as "spatially connected" is gained. "However, all points are not equal: People tend to give landmarks and more "important" places in the network a special status. Increasing familiarity with the environment leads to learning routes, which connect the locations" (see also Couclelis *et al.* 1987,

Golledge 1992). Route knowledge has a "directional bias" in "the early stages of spatial understanding," but upon further "recall and evaluation" the routes are perceived as a connected network rather than in isolation. "Further familiarity leads to survey or map knowledge in which locations are linked efficiently and spatial relations such as distance and orientation are abstracted," (Gluck 1991, p.119).

There are many areas of individual and collective choice behaviour that employ the concept of a hierarchy to represent the various levels of decision making (e.g., with "nested" discrete choice, logit methods.) While the concept of hierarchy is commonly used, the AHP offers the operational advantage of a discrete method particularly useful in situations involving both tangibles and intangibles. Above all, at the core of the AHP is a method of paired comparisons and a measurement scale that is derived directly from the interpretation of data. If cognitive maps are constructed through a process of interpretation, experience, and learning, then what better method is there than one with a psychometric measurement scale that corresponds directly to the tasks performed by subjects.

3. THE IMAGE OF A UNIVERSITY CAMPUS: A COGNITIVE MAPP APPLICATION OF THE AHP

3.1 Methodology and Data

Lynch (1960) developed a method of surveying, mapping, and classifying city images with his seminal case studies of the core area of three U.S. cities. As Golledge and Stimson (1997, p. 250) note recently, the image of the city (Lynch 1960) "not only served to focus attention on perceptual and cognitive qualities of the urban environments, but it also provided a conceptual framework for the discussion of the structural components of city images that still occupies a primary place in the literature of city structure." Lynch's taxonomy was principally concerned with the legibility or imageability of spatial form of the city, though it "has been the basis for all wayfinding discussions since its presentation" (Gluck 1991, p. 120). Lynch (1960) used five elements as a "convenient" classification of public images of physical form. The well-known elements are paths, edges, districts, nodes, and landmarks. The imageability or legibility of physical form was determined from the observation of the elements and their interrelations.

The survey methodology consisted of a "lengthy" public interview (small sample) as well as a systematic field survey by "trained" observers familiar with the concept of imageability. The field survey was done independently of the public interview. "Surprisingly", Lynch (1960, p. 144) noted, "the field analyses in Boston and Los Angeles" rendered "accurate predications of the images derived from the verbal interview material." The field analysis in Jersey City (which Lynch characterised as "poorly differentiated") predicted nearly "two-third of the interview image," with the majority of the major elements identified in both field analysis and the interview material. In all three case studies Lynch reported the high consistency in "the relative ranking of elements" (Lynch 1960, p. 144). The results suggested a plausible technique for visual analysis that could test hypotheses and predict public image(s) of a place.

While the objective of testing hypotheses and of predicting public image(s) from a method of visual analysis of a place were evidently attained, Lynch raised concern regarding its certain limitations, namely that the method emphasised "single elements, and under-emphasised their patterning into a complex visual whole" (Lynch 1960, p. 144). Considered in pair (e.g. paths-nodes, paths-districts), Lynch noted, the elements could strengthen or weaken, "reinforce" or "destroy" each other. Lynch added, "While data on single elements and elements types was perhaps adequate, there was a lack of information on element interrelation, patterns, sequences, and wholes. Better methods must be evolved to approach these vital aspects." (Lynch 1960, p. 155; see also Lynch 1990)

We used an approach with a method that aids in the measurement of the interrelations of Lynch's elements of a visual survey—the method of paired comparisons, which is at the kernel of the AHP procedure, scale, and the calculus of consistency. Formally, the paired comparisons of the (five) elements are performed by means of Saaty's square, reciprocal matrix with unit rank whose eigenvector solution gives the relative importance, or dominance, of the elements on a ratio scale. The inputs to such a matrix are provided by means of a questionnaire, which solicits each field surveyor's perception of the relative dominance of the elements within each pair by using the AHP nine-point scale [1,9]. The surveyors were graduate students enrolled in a city planning course in which Lynch (1960) was one of the required texts. The main campus of the university, a site approximate in size (1-1/2 x 2-1/2 miles) to the areas in the case studies originally studied by Lynch (1960) was selected for a study of imageability. Subsequently, they were given a questionnaire to enter individual assessments and comments regarding legibility of the elements as observed in the field.

The relative importance and rankings of the elements are derived from the structure of the elements' interrelations instead of the number of times a single element is observed. Furthermore, from the consistency (index) of each subject's assessments of the element interrelations we obtain an account of the perceived connectivity of the elements. Thus, an account of the image of structure is provided, critical information of how subjects view typological organisation, and helpful in design decisions affecting legibility of physical form as a whole. This approach retains the essentially qualitative conceptual property of Lynch's near-classic methodology while it allows for statistical, quantitative account of observer variation and group consensus of the image(s) of structure.

The essential question for each subject to pose is this: In considering each pair of elements, which element has a greater weight, i.e., has greater dominance or importance, compared to the other in contributing to the legibility of a place as seen in the field? One subject's (value) judgements of the relative importance or dominance of the elements to the legibility of the place (described later in detail) are given below in matrix **A**. The comments of the subject whose value judgements are entered in **A** were stated thus:

A =

	Path	Edge	District	Node	Landmark	Weight
Path	1	3	3	1	2	0.219
Edge	1/3	1	1	1/4	1/4	0.071
District	1/3	1	1	4	4	0.071
Node	1	4	1/4	3	3	0.372
Landmark	1/2	4	1/4	1/3	1	0.266

For example, in the comparison of paths and districts: "Since so much of the campus is encompassed by the major district, paths become moderately more important [a_{13} = 3]". With regard to paths and nodes, "As they should be, paths and nodes are connected on this site. The clarity of one decides the clarity of the other" (a_{14} = 1). The value entered in the comparison of edges and nodes (a_{24} = 1/4) was justified thus: "Especially in a campus setting, nodes act as a resource to people. Information can be had by going there. Edges help people know where they are, but nodes tell them how to get where they are going." And so on.

The relative importance of the elements (determined by the principal eigenvector of **A**) to the legibility of the site as seen by the subject (#5), in a descending order, is as follows: nodes (37.2%), landmarks (26.6%), paths (21.9%), and edges or districts (7.01%), which are shown in a column next to **A**. Was this subject logically consistent in performing pairwise comparisons of the elements, or were the values entered randomly? Saaty (1980) provides a calculus for determining consistency.

For the judgements entered in **A**, λ_{max} = 5.26112, and with the number of elements n = 5, we get the value of the consistency index CI = (5.26112 – 5)/(5 –1) = 0.0653. Its random value for a matrix of order 5 is 1.12 (Saaty 1980). It turns out that the judgements regarding paired comparisons of the elements made by the subject (#5), indicated by the consistency ratio, *CR* = 0.0653/1.12 = 0.058, are acceptable if we use a limit suggested by Saaty (1980) that a value of less than 10% indicates good consistency. If the value had exceeded this benchmark, judgements are revised so as to improve upon logical consistency. And thereby the method encourages further information and learning with observation and reflection.

The relative weights of the elements (derived from eight subjects)—constrained within an acceptable level of consistency (*CR*<10%)—are shown in Table 3. Also shown are the (ordinal) rankings of the elements (in parentheses). The mean weight (and rank) of the elements for the subjects as a group is shown in the last column of Table 3.

Table 3. The relative importance of the elements with rankings from a sample of eight subjects with consistency ratios < 10%.

	Subjects								
Elements	1	2	3	4	5	6	7	8	Mean
Paths	0.513	0.130	0.408	0.261	0.219	0.490	0.283	0.152	0.307
	(1)	(3)	(1)	(2)	(3)	(1)	(2)	(3)	(1)
Edges	0.261	0.062	0.260	0.086	0.071	0.164	0.033	0.262	0.150
	(2)	(5)	(2)	(5)	(4)	(3)	(5)	(2)	(5)
Districts	0.129	0.227	0.083	0.151	0.071	0.085	0.068	0.445	0.157
	(3)	(2)	(4)	(4)	(4)	(4)	(4)	(1)	(4)
Nodes	0.063	0.495	0.083	0.319	0.372	0.049	0.164	0.089	0.204
	(4)	(1)	(4)	(1)	(1)	(5)	(3)	(4)	(2)
Landmarks	0.033	0.085	0.166	0.183	0.266	0.213	0.453	0.052	0.181
	(5)	(4)	(3)	(3)	(2)	(2)	(1)	(5)	(3)
Consistency(%)	5.3	6.0	7.8	4.1	5.8	7.6	6.3	0.6	

In addition to gauging the consistency of individual responses, Table 3 indicates the agreement (or disagreement) among the subjects in the perception of the relative importance of the elements. Kendall's coefficient of concordance given by the value of W = 0.1495 indicates a weak agreement among the subjects ($0 \leq W \leq 1$, with zero as perfect disagreement, and one as perfect agreement). However, a problem of statistical discernability is posed with the corresponding p = 0.3274, due to the small size of this sample. The limitation of a small sample notwithstanding, reliability analysis (ANOVA) indicates that, on balance, the subjects' ratings (using relative weights as data), or rankings (using ordinal ranks) of the elements are similar. So are the mean ratings (p = 0.2733), or rankings (p = 0.3274), of the elements.

4. DISCUSSION

Lynch (1960) viewed his study methodology to be further refined through development of methods which aid in the investigation of the interrelation of "unlike elements" in order to discern "patterns, sequences, and wholes" (Lynch 1960, p. 155), after the identification of the parts (elements). Wayfinding research, for example, indicates that landmarks and routes are learned "conjointly and very quickly," rather than considered separately (Blades 1991, p. 158). As Blades (1991) notes, examination of the relative importance of environmental features like landmarks and routes as well as of the level of experience required in wayfinding is a research focus.

The matrix method used in this study aids in an investigation of the interrelation of the elements. How the "elements operate together" (Lynch 1960, p. 84), as viewed by a group of subjects, is determined in context. The context for this study was the main campus of an urban university. The concept of a "campus" ideally connotes a whole (district) determined, "reinforced," or enriched by the interrelation of comprising parts (paths, nodes, landmarks, and edges). Hence, the campus provides an ideal (as well as a manageable survey) setting where the concept of structural imageability, i.e., the legibility of the (campus) physical form by the pattern or structure of the interrelations among the five elements could be examined.

While maintaining logical consistency in the paired comparisons of the elements, the subjects' perceptions of the degree of interrelation of the elements vary, reflected in the relative weights (or rank) of the elements (Table 3, above). The variation in the perception of the relative dominance of the elements by subjects is in effect a disagreement on structure or pattern of the interrelations of the elements, which suggests a problem of (lack of) imageability or legibility of the physical form of the campus. Problems identified in the visual survey of the campus may be regarded as factors contributing to a lack of agreement on a "legible" or coherent structure (see Figure 3a).

An overview of the problems of the campus image as surveyed by the subjects is given in Figure 3a and contrasted (explained) with the observation of the elements and their perceived relations in Figure 3b. As an exercise in cognitive mapping, experience, learning, and reflection play a role. Hence also the facility with which the problems of the image by the expert subjects are identified. The contrast of the two figures, furthermore, gives an indication of the learning process with an initial "sketch" of the problems that was followed later by a more refined, comprehensive, and even cartographically accurate representation.

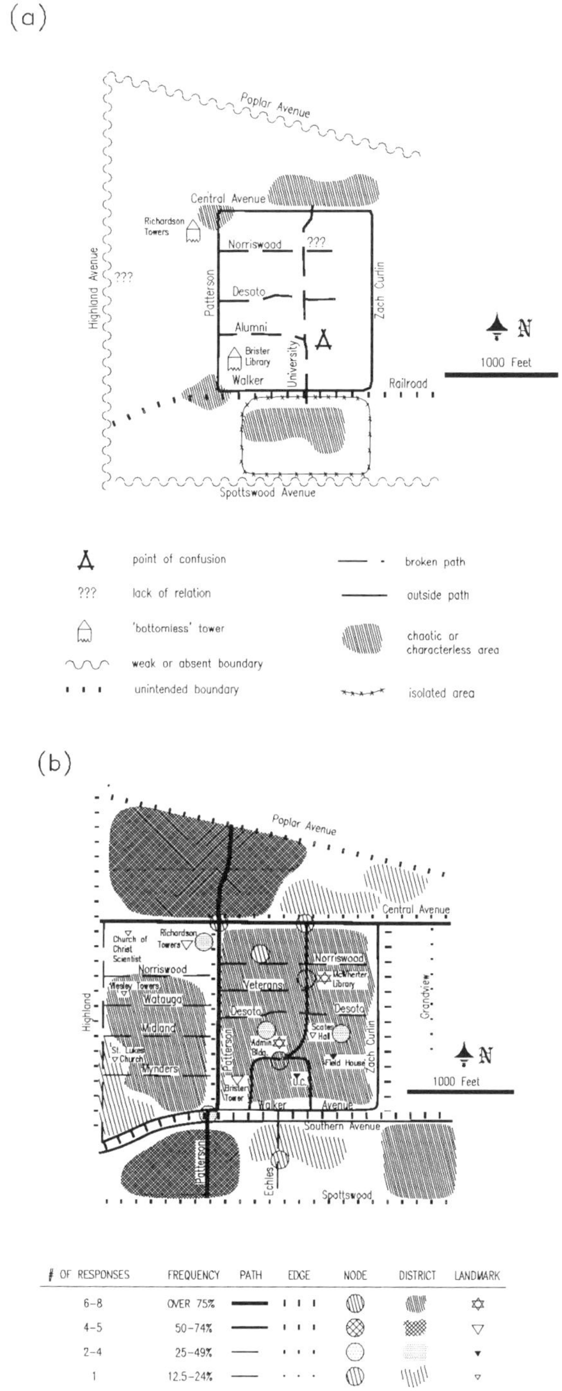

Figure 3. (a) Problems with the campus image; (b) The image as derived from group survey.

For example, while paths are dominant in the image(s) of the subject (mean weight (rank): 0.3047(1)), they are weakly related to other elements such as nodes or landmarks. The relative dominance of the (grid) paths does indicate their contribution to the clarity or legibility of the metropolitan organisation (globally) in which the campus is located, but their contribution as a "resource in organisation" (Lynch 1960, p. 84) of the campus is limited (locally) due to the lack of the relations of paths to other elements. The dominance of paths, however, is functionally reflective of the commuter orientation of a campus in an urban setting, rather than one which is "self contained," particularly with respect to housing. Therefore, paths are most likely features to be experienced early and with a high priority before knowledge of the overall spatial configuration in relation to other environmental features is gained.

The image of districts, with a relatively low mean weight (rank) of 0.157 (4), is subordinated by the dominant image of the paths. The image is evidently reflective of the inadequate "concentration and repetition of themes" (Lynch 1960, p. 165), whether natural or the built-environment, which if otherwise were present they could contribute to a stronger identification of the campus as a district, upon entrance or departure by an observer. The railroad lines which fragment the campus and conflict with pedestrian crossing, and thoroughfares (also considered as paths) which pose a treacherous vehicular vs. pedestrian movement are seen as weak edges (mean weight (rank) = 0.150 (5)). In contrast, nodes and landmarks are stronger (mean weight (rank) = 0.204 (2) and 0.181 (3), respectively), the image(s) of which are reinforced by the recent addition of a new library (landmark) fronting a landscaped pedestrian space (node) at the core of the campus.

We have applied our methodology diagnostically to determine how environmental structure is perceived by means of the relative dominance of its interrelating component parts. But we also see a place for our methodology in the determination of a "visual plan" (see Lynch 1960) in which the paired comparisons and ratings of the structural elements are performed normatively. Such comparisons consider the issues of how the elements should be given priorities when considered jointly in order to strengthen the legibility or visibility of physical form as a whole (see also Banai 1999).

5. CONCLUSIONS

From Lynch's seminal study of city structure to the recent investigations of wayfinding, mental representation of space is characterised by processes of observation, interpretation, experience, and learning. The AHP exhibits the facility to model such processes because it provides a comprehensive, scientific theory that accounts for people's subjective sensation of and response to objective stimuli. Ironically, a theory of how both the physical and social world is mentally represented (see Saaty, 1998) has received little attention in environmental cognition literature and lagged behind other multidisciplinary areas of application and development. It turns out, however, the AHP holds in common certain key conceptual and methodological areas addressed in the cognitive mapping and modelling research while it provides a plausible, alternative paradigm for environmental cognition. The relative measurement scale of the AHP is particularly useful in the analysis of the qualitative, topological properties of spatial organisation. Paired comparisons of intangible, physical form qualities are all that we can opt for in the absence of a scale with a unit (Saaty, 1998). A promising future application is to probe the distinction, and the chasm, between two types or "levels" of common vs. expert knowledge in spatial cognition (see Golledge 1993). Saaty's AHP in this context is helpful as a technique for multivariate group or public decision making and consensus building. Computer-aided, interactive survey protocol thus potentially serves as a means for not only the scanning of the public image(s), but also as a means for public participation, group deliberation and preference assessment toward both identification and design(s) of desirable image(s) of a place.

REFERENCES

BANAI, R. 1993. Fuzziness in geographical information systems: Contributions from the analytic hierarchy process. International Journal of Geographical Information Systems 7:315-329.

BANAI, R. 1998. Transit-oriented development suitability analysis by the analytic hierarchy process and a geographic information system: A prototype procedure. Journal of Public Transportation 2: 43-65.

BANAI, R. 1999. A methodology for The Image of the City. Environment and Planning B 26: 133-144.

BANAI-KASHANI, R. 1984. Travel demand (modal split) estimation by hierarchical measurement. Journal of Advanced Transportation 18:37-54.

BANAI-KASHANI, R. 1990. Discrete mode-choice analysis of urban travel demand by the analytic hierarchy process. Transportation 16:81-96.

BHASKAR, R. 1989. Reclaiming Reality: A Critical Introduction to Contemporary Philosophy Verso, London.

BLADES, M. 1991. Wayfinding theory and research: The need for a new approach. Pages 137-165 in Mark D. M. and Frank A.U. (eds.) Cognitive and Linguistic Aspects of Geographic Space. Kluwer Academic Publishers, Dordrect.

COUCLELIS, H., GOLLEDGE, R.G., GALE, N. AND TOBLER, W. 1987. Exploring the anchor-point hypothesis of spatial cognition. Journal of Environmental Psychology. 7:99-122

FRANK, A. U. 1991. Qualitative spatial reasoning about cardinal directions. Autocarto 10: 148-167.

GARLING, T. AND GOLLEDGE, R. G. 1989. Environmental perception and cognition. Pages 203-236 in Zube E.H. and Moore G.T. (ed.) Advances in Environmental Behaviour and Design 2: Plenum Press, N.Y.

GARLING, T. AND GOLLEDGE, R. G. 1993 (ed.) Behavior and Environment: Psychological and Geographical Approaches. Elsevier, New York.

GARLING, T., KALEN, T., ROMANUS, J. AND SELART, M. 1998. Computer simulation of household scheduling. Environment and Planning 30:665-679.

GOLLEDGE, R.G. 1992. Place recognition and way finding: Making sense of space Geoforum 23:199-214.

GOLLEDGE, R. G. 1993. Geographical perspectives on spatial cognition. Pages 16-46 in Garling T. and Golledge R.G. (eds.) Behaviour and Environment: Psychological and Geographical Approaches. Elsevier Science Publishers, NY.

GOLLEDGE, R. G. AND STIMSON, R. J. 1997. Spatial Behaviour: A Geographic Perspective The Guilford Press, New York.

GLUCK, M. 1991. Making sense of human wayfinding: Review of cognitive and linguistic knowledge for personal navigation with a new research direction. Pages 117-135 in Mark D. M. and Frank A.U. (eds.) Cognitive and Linguistic Aspects of Geographic Space Kluwer Academic Publishers, Dordrect.

HIRTLE, S.C. AND JONIDES, J. 1985. Evidence of hierarchies in cognitive maps. Memory and Cognition 13:208-217.

KITCHIN, R.M. 1994. Cognitive maps: What are they and why study them. Journal of Environmental Psychology 14:1-19.

LINDBLOM, C. E. 1959. The science of "muddling through". Public Administration Review 19,2: 79-88

LYNCH, K. 1960. The Image of the City. MIT Press, Cambridge, MA.

LYNCH, K. 1990. Reconsidering The Image of the City (1985) Pages 247-56 in Banerjee T.and Southworth M. City Sense and City Design. (Eds.) MIT Press, Cambridge, MA.

MCDONALD, T. AND PELLEGRINO, J. 1993 Psychological perspectives on spatial cognition. Pages 47-82 in Garling T. and Golledge R.G. (eds.) Behavior and Environment: Psychological and Geographical Approaches. Elsevier Science Publishers, NY.

MCNAMARA, T.P. 1986. Mental representations of spatial relation. Cognitive Psychology 18:87-121.

MCNAMARA, T.P. 1991. Memory's view of space. Pages 147-186 in Bower G.H. (ed.) The Psychology of Learning and Motivation: Advances in Research and Theory. Academic Press, NY.

MCNAMARA, T.P. 1992. Spatial representation. Geoforum 23: 139-150.

MCNAMARA, T., HARDY, J. AND HIRTLE, S. 1989. ive hierarchies in spatial memory. Journal of Experimental Psychology: Learning, Memory and Cognition 15:211-227.

MALCZEWSKI, J. 1996. A GIS-based approach to multiple criteria group decision making. International Journal of Geographical Information Systems 10:955-971.

MILLER, G.A. 1956. The magical number seven plus or minus two: Some limits on our capacity for processing information. Psychological Review. 63:81-97.

MONTELLO, D.R. 1991. The measurement of cognitive distance: Methods and constructvalidity. Journal of Environmental Psychology 11:101-122.

MONTELLO, D.R. AND FRANK, A.U. 1996. Modelling directional knowledge and reasoning in environmental space: testing qualitative metrics. Pages 321-344 in Portugali J. (ed.) The Construction of Cognitive Maps. Kluwer, Netherlands..

PERUCH, P., GIRUDO, M.D. AND GARLING, T. 1989. Distance cognition by taxi drives and the general public. Journal of Environmental Psychology 9:233-239.

SAATY, T.L. 1978. Exploring the interface between hierarchies, multiples objectives and fuzzy sets. Fuzzy Sets and Systems 1:57-68.

SAATY, T.L. 1980. The Analytic Hierarchy Process. McGraw-Hill, New York.

SAATY, T.L. 1986. Axiomatic foundation of the analytic hierarchy process. Management Science 32, 7:841-855.

SAATY, T.L. 1990. Physics as a decision theory. European Journal of Operational Research 48:98-104.

SAATY, T.L. 1995. Transport planning with multiple criteria: The analytic hierarchy process applications and progress review. Journal of Advanced Transportation 29:81-86.

SAATY, T.L. 1996. Multicriteria Decision Making. RWS Publications, Pittsburgh, Pa.

SAATY, T.L. 1998. Reflections and projections on creativity in operations research and management: A pressing need for a shift in paradigm. Operations Research 46:9-16.

SIMON, H.A. 1981. The Sciences of the Artificial. MIT Press, Cambridge, MA.

SIMON, H.A. 1982. Models of Bounded Rationality. MIT Press, Cambridge, MA.

SIMON, H.A. 1983. Reason in Human Affairs. Stanford University Press, Stanford, CA.

SMITH, T.R., PELLEGRINO, J.W. AND GOLLEGE, R. 1982 Computational process modeling of spatial cognition and behavior. Geographical Analysis 14:305-325.

ZADEH, L. A. 1990. The birth and evolution of fuzzy logic. International Journal of General Systems 17:95-105.

Chapter 11

Potential Allowable Cut of Finland Using the AHP to Model Landowners' Strategic Decision Making

Mauno Pesonen
Finnish Forest Research Institute, Finland

Key words: Nonindustrial private forest landowner, potential allowable cut, timber management strategy, utility function.

Abstract: This paper analyses the potential allowable cut of Finland based on nonindustrial private forest (NIPF) landowners' choices as to timber management strategies. Alternative timber management strategies were generated, and the choices and factors affecting the preference of strategies by NIPF landowners were studied. Timber management strategy choices were determined by maximising the utility functions of the NIPF landowners. The parameters of the utility functions were estimated using the analytic hierarchy process (AHP). The level of the potential allowable cut was compared to cutting budgets based on the 7th and 8th national forest inventories (NFI7 and NFI8), to combined private forestry plans, and to the actual harvest from nonindustrial private forests. The potential allowable cut was calculated using the same MELA system as has been used in the calculation of the national cutting budget. The potential allowable cut defined in the study was 19% higher than the average of the actual harvest during the years 1989-1994. Correspondingly, the potential allowable cut defined in the present study was 13% lower than the NFI8-based greatest sustained allowable cut for the 1990s. Using the method presented in this study, which is based on choices of timber management strategies, regular cutting budgets can be calculated more realistically than before.

1. INTRODUCTION

1.1 The Sustained Allowable Cut in Finland

According to Sevola (1998), there are 437,000 nonindustrial private forest (NIPF) holdings in Finland, covering 62% of the country's forest area.

D.L. Schmoldt et al. (eds.),
The Analytic Hierarchy Process in Natural Resource and Environmental Decision Making, 167–185.

These NIPF forests represent 70% of the growing stock volume and 73% of the annual increment. Roundwood removals from NIPF holdings have amounted to 75-85% of the total roundwood removals during the past ten years (Sevola 1998). Commercial cuttings from NIPF lands account for 50-70% of the total amount of timber used by the country's forest-based industries. Thus, NIPF forestry and timber supply has a very prominent role in the forest sector and in the Finland's national economy.

Generally, the allowable cut has been viewed from the perspective of many factors and sources of information. Cutting budgets utilising forest inventory data provided by national forest inventories (NFI) and cutting budgets obtained by combining nonindustrial, private forestry plans are the major sources of information for formulating the national allowable cut. The two major weaknesses related to cutting budgets based on NFI data are: (1) national cutting budgets have been calculated assuming that all the country's forests are treated as a single forestry unit, and (2) the variability of goals subscribed to by NIPF landowners has been ignored. Due to differences between inventory data and due to the calculation method, cutting budgets derived by combining NIPF forestry plans can be more than 30% lower than those based on NFI data (Forest 2000 Program 1985). Thus, treating forest resources simply as one entity leads to overestimating the allowable cut. In fact, the cutting budgets derived by combining NIPF forestry plans can be more than 30% lower than those based on NFI data (FOREST 2000... 1985). On the other hand, forestry plans made for NIPF holdings are often deliberate underestimations of the actual cutting potential; the cutting budgets presented in the associated forestry plans can be nearly 20% smaller than the actual allowable cut based on sustained forestry (Pesonen and Räsänen 1993).

1.2 Strategic Decisions in Managing NIPFs

The goal in strategic forest planning is to define the general strategy for the management of a forest holding, and to maximise the forest owner's utility by allocating the resources according to the owner's goals (Kajanus *et al.* 1998). At the strategic planning stage, it is important to create a broad view of the decision landscape. In order to maximise the utility of the forest owner, a strategic analysis should be carried out at the enterprise level. This includes determining production possibilities for the forest resource in a manner comparable with the enterprise's other lines of business, e.g. in terms of income, costs, value of assets, and working hours.

Strategic planning operates on future production possibilities; the starting point for which is the convertibility of the production factors and their allocation (e.g., Kast and Rosenzweig 1974). When applied to NIPF management planning, a strategic view includes the production of alternative, strategic-level programs for timber production and silviculture. Timber management covers a range of strategies from no cuttings at all to maximum cuttings within the limits of timber production. For instance, timber management strategies can be described by the intensity and the recurrence of cuttings.

Most NIPF landowners have long-term perspectives and strategic views concerning forest management (Lönnstedt 1989). It is important to understand the strategic decisions of NIPF landowners for several reasons: (1) when predicting the timber supply from these private forests for future investments by forest industries (Lönnstedt and Roos 1993), and (2) when planning governmental forest policy in general. In Finland, present-day NIPF management planning is basically tactical. Landowners lack information about actual, strategic-level decision alternatives and their consequences. Furthermore, decision analysis (i.e., giving recommendations about decisions and making decisions) is often separated in planning. The importance of planning in the production of decision alternatives, and in defining landowners' preferences, is often ignored.

While strategic forest management planning is lacking in Finnish NIPF forestry, landowners tend to underestimate their allowable cut. Furthermore, forestry plans are usually underestimates when compared to the sustained allowable cut of forest holdings (Pesonen and Räsänen 1993). Moreover, 60% of the landowners have actually harvested even less than the cutting budget presented in forestry plans (Pesonen *et al.* 1994).

Many studies on strategic forest management planning (e.g. Wardle 1965, Kilkki 1968, Ware and Clutter 1971, Kangas and Pukkala 1992, Kajanus *et al.* 1998) have been done and several tools (Siitonen 1983, Johnson and Jones 1986, Jonsson *et al.* 1993, Pukkala and Kangas 1993) have been developed for strategic forest management planning. However, few studies have been conducted concerning the regional cutting budgets derived from the strategic goals of NIPF landowners (Pesonen 1995, 1996). Lönnstedt and Roos (1993) concluded that the cutting potential based on objectives of NIPF landowners ensures an adequate supply of wood raw material for future investments by Sweden's forest-based industries.

1.3 Modelling Strategic Decision Making of NIPF Landowners

Modelling the strategic decision making of NIPF landowners, like any other attempt at modelling human behaviour, is a complex and multidimensional task. Only a few studies have been conducted on the strategic decisions of NIPF management (Lönnstedt and Törnqvist 1990,

Hansson *et al.* 1990, Pukkala and Kangas 1993) and the factors affecting these decisions have received little attention.

One of the methods used in decision analysis is the analytic hierarchy process (AHP). The AHP has been applied to wide variety of decision situations. Moreover, there have been studies on the applications of the AHP to forest management planning (Mendoza and Sprouse 1989, Kangas 1992, Kangas and Pukkala 1992, Kurttila *et al.* 1998). The AHP is a mathematical method for analysing complex decision problems with multiple criteria (Saaty 1977, 1980). Basically, the AHP is a general theory based on certain mathematical and psychological foundations. When applying the AHP, a hierarchical decision schema is constructed by decomposing the decision problem into its decisions elements: goals, objectives, attributes, and decision alternatives. The relative importance or preference of the decision elements at each level are compared in a pairwise manner with regard to the element preceding them in the hierarchy. In this study, the AHP was used to determine NIPF landowners' choices of preferred timber management strategies.

1.4 Aims of the Study

The aims of this study were to (1) generate alternative timber management strategies for NIPF landowners, (2) determine their preferences for alternative timber management strategies, and (3) calculate the potential allowable cut from nonindustrial private forests in Finland. The potential allowable cut is calculated using TASO[1] and NFI data to show how reliable it is to generate regional allowable cuts with TASO data.

In this study, *timber management strategy* is defined as an alternative for a NIPF landowner in the utilisation of his/her forest property and *potential allowable cut* means the regional cutting budget calculated for a particular forestry region derived from the landowner's choices of timber management strategies. Choices of timber management strategies are determined by maximising the utility functions of NIPF landowners. Parameters of the utility function are estimated using the AHP. The potential allowable cut is derived from the NIPF landowners' choices of timber management strategies, and it is compared to (1) cutting budgets based on NFI7 and NFI8 data, (2) combined NIPF management plans, and (3) actual harvest (1989-1994) on NIPF lands. The potential allowable cut is calculated using the

[1] TASO is the Finnish forest planning system, which has been used in non-industrial private forestry since 1987. A TASO forest plan consists of compartment-level data, compartment maps and management suggestions.

same MELA system (Siitonen 1983) as has been used in the calculation of national cutting budgets in Finland.

2. THE UTILITY FUNCTION AND TIMBER MANAGEMENT STRATEGIES

2.1 Estimating Utility Function Parameters Using the AHP

According to a generally accepted economic theory, rational decision makers (such as forest owners) are supposed to maximise their utility when they make decisions (e.g., Hirshcleifer 1984). Theoretical.y, then, the preferences of a decision maker are often modelled as a function called the *utility function*. Utility theory has been further developed to solve decision problems with multiple objectives in complex decision situations, i.e. multi-attribute utility theory (e.g., von Winterfeldt and Edwards 1986, Kangas 1992, Mykkänen 1994).

The linear and additive utility function applied in this study has been the one most commonly used. It is also considered to be the easiest to interpret (Pukkala and Kangas 1993). In the formulation of the utility function for determining the choice of timber management strategy, the overall utility obtained from the use of forest property consists of the utility obtained from the economic and the non-economic benefits of the forest property. In this study, the *economic benefits* consist of the utility of timber production and the *non-economic benefits* of other benefits. Therefore, the form of the additive utility function (Pesonen 1996) is:

$$max\, U(S_p) = a_1 u_{econ}(S_p) + a_2 u_{non}(S_p) \tag{2.1}$$

where

U total utility obtained from use of forest property (i.e. utility from timber management strategy)

$u_{econ}(S_p)$ utility obtained from economic benefits of timber management strategy

$u_{non}(S_p)$ utility obtained from non-economic benefits of timber management strategy

S_p timber management strategy, $1<p<m$

m number of timber management strategy

a_1, a_2 parameters describing importance of respective criterion

In pairwise comparisons, landowners decided which one of the two timber management strategies they preferred, both with respect to the economic and non-economic benefits from their forest property. Landowners were able to express their preference as equal (1), slight (3), moderate (5), strongly demonstrated (7), or absolutely preferred (9), or using intermediate values (2, 4, 6, 8).

2.2 The Definition of Timber Management Strategies

To solve the parameters of the utility function (2.1), five alternative timber management strategies were computed using the MELA system. MELA is a Finnish LP-based system for long-term timber management planning (Kilkki and Siitonen 1976, Siitonen 1983, 1993). Strategies based on TASO data calculations were described for each landowner with the objective and constraint variables derived from the MELA parameters (Figure 1). The planning horizon was 20 years, divided into four 5-year periods. In the calculations, the forest-holding-level development of several forest characteristics was described and illustrated for the landowners.

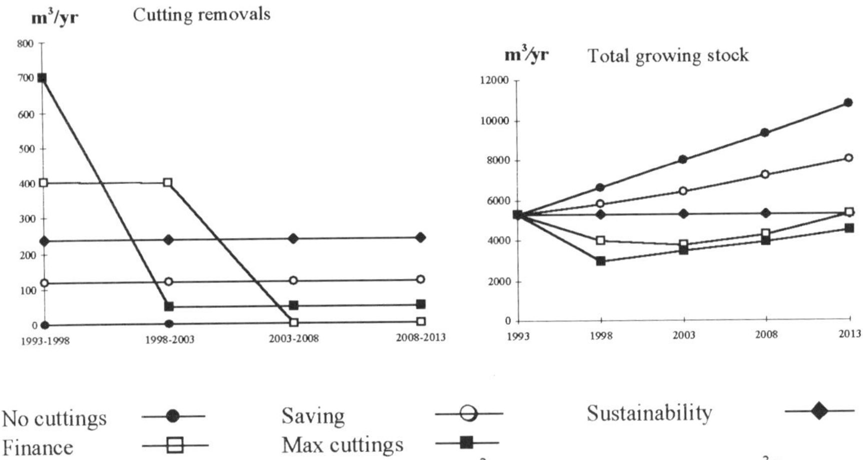

Forest area: 50.8 ha Initial growing stock: 5332 m^3 Mean initial volume: 105 m^3/ha
Average of the realised commercial cuttings during 1988-1992: 347 m^3/a.

Figure 1. Alternative timber management strategies described as the pattern of removals and total growing stock during the planning period (an example of calculations made for each NIPF landowner, representing a sample case of the forest holdings).

Each landowner was provided with five alternative timber management strategies covering a planning horizon of 20 years. In principle, the main differences between the strategies can be described in terms of intensity and the recurrence of removals. The objective variable used in optimisations was the maximisation of the stumpage earnings for the first planning period (the constraints for each strategy are presented below). The applied timber management strategies were as follows:

S_1 "NO CUTTINGS"

- removals set to zero

S_2 "SAVING"

- removals set to half of the removals under the condition "SUSTAINABILITY"

S_3 "SUSTAINABILITY"

- even flow of removals over the planning horizon
- even flow of stumpage earnings over the planning horizon
- even amount of clear-cut areas over the planning horizon
- volume of sawtimber equal to, or greater, than at the beginning of horizon
- market value of growing stock at the end of planning horizon being at least the same as at the beginning

S_4 "FINANCE"

- even flow of removals during the first two planning periods
- market value of the growing stock at the end of the planning horizon being at least the same as at the beginning

S_5 "MAX CUTTINGS"

- even flow of removals during the last three planning periods

NIPF landowners were asked to prioritise the timber management strategies according to their personal goals and preferences for their forest. First, the NIPF landowners were asked to compare the importance of the economic and non-economic benefits of their forest holdings. Second, pairwise comparisons were made between the management strategies, considering the economic and the non-economic benefits separately (Figure 2). The AHP process resulted in the relative priorities for each strategy being scaled 0-1. For each landowner, the strategy with the highest global priority (i.e., one that maximises the overall benefit) represented the preferred alternative.

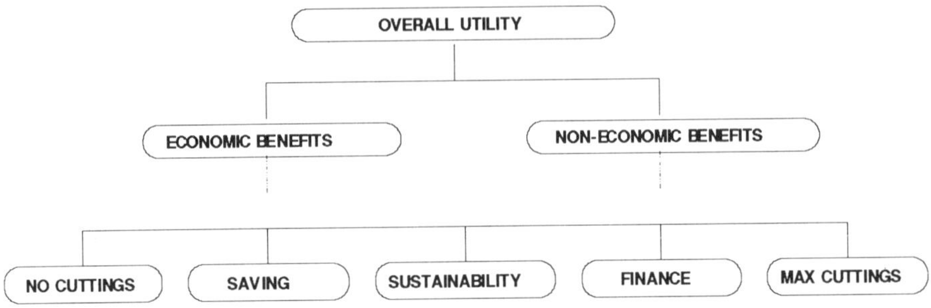

Figure 2. Decision hierarchy when selecting the preferred timber management strategy.

3. MATERIALS AND METHODS

Ownership data consisted of those NIPF holdings in southern Finland that had developed a forestry plan during the years 1989-1993. The data were collected from the TASO planning system, which has used for forest management planning on nonindustrial private forests since 1987 (Ranta 1991). Data were collected on 66,706 forest holdings with a total forest area of 2,882,114 ha and an average holding size of 43.2 ha (Table 1a). According to the official register on Finnish farm holdings (Pihljerta 1994), the forest area in southern Finland was 8,466,100 ha (Table 1b). In comparing those data, small forest holdings were under-represented and large forest holdings were over-represented.

Forest holdings were divided into four groups according to their forest area: 5-19.9, 20-49.9, 50-99.9, and over 100 hectares. Stratified sampling was carried out according to forest holding size, so that the number of holdings in each sample group was determined by assigning a 4% maximum standard error in the initial volume (m^3/ha) within the groups. The sample consisted of 4,105 forest holdings giving a combined forest area of 214,662 ha and an average holding size of 52.3 ha (Table 1c).

After sampling, background information on NIPF landowners, their forest property and forestry goals were collected by means of a two-phase mail questionnaire. Two phases were necessary because landowners had to be asked in advance for their permission to use the data from their forestry plans. In the first phase, the landowners were asked some questions about their ownership characteristics, economy, and educational background.

Table 1a. Distribution of the basic data according to forest holding size.

Size of holding (ha)	Forest area (Hectares)	%	Number	%	Average size
5-19.9	291,754	10.1	23,320	35.0	12.5
20 - 49.9	863,727	30.0	26,735	40.1	32.3
50 - 99.9	795,919	27.6	11,598	17.4	68.6
100 -	930,714	32.3	5,053	7.6	184.2
Total	2,882,114	100.0	66,706	100.0	43.2

Table 1b. Forest area distribution according to the official register on Finnish farm holdings.

Size of holding (ha)	Forest area (Hectares)	%	Number	%
5-19.9	1,483,000	17.5	123,696	50.1
20 - 49.9	2,962,225	35.0	84,635	34.3
50 - 99.9	2,174,325	25.7	28,991	11.7
100 -	1,846,550	21.8	9,487	3.8
Total	8,466,100	100.0	246,809	100.0

Table 1c. Distribution of the sample according to forest holding size.

Size of holding (ha)	Forest area (Hectares)	%	Number	%	Average size
5-19.9	16,763	7.8	1,393	33.9	12.0
20 - 49.9	37,281	17.4	1,157	28.2	32.2
50 - 99.9	64,755	30.2	927	22.6	69.9
100 -	95,863	44.7	628	15.3	152.6
Total	214,662	100.0	4,105	100.0	52.3

Table 1d. Distribution of the sample following mail questionnaire according forest holding size.

Size of holding (ha)	Forest area (Hectares)	%	Number	%	Average size
5-19.9	5,459	7.0	439	32.1	12.4
20 - 49.9	11,660	14.9	358	26.2	32.6
50 - 99.9	21,502	27.5	310	22.7	69.4
100 -	39,597	50.6	260	19.0	152.3
Total	78,218	100.0	1,367	100.0	57.2

The second phase of the questionnaire was preceded by calculations, which were presented to each landowner as alternative timber management strategies. In the second phase of the questionnaire, a total of 1,367 acceptable responses were received representing an average holding size of 57.2 ha (Table 1d), which was larger than the average of the entire population. Data collection during the two phases of the questionnaire study resulted in a situation in which small holdings (under 20 ha and 20-50 ha) were under-represented and large holdings (over 100 ha) were over-represented when compared to the corresponding proportions in the Official

Register of Finnish Farms (Pihljerta 1994). There are least two reasons causing this bias: (1) forest holdings with forestry plans are generally above-average in size, and 2) presumably, landowners with large forest holdings were more interested in study participation. Due to this bias, all the results (distribution of choices of timber management strategies, potential allowable cut) were weighted with the area group distribution of the Official Record of Finnish Farms (Pihljerta 1994).

The reference material for the calculations consisted of the cutting budgets calculated on the basis of sample plot data provided by NFI8 (hereafter referred as the NFI data). The same timber management strategies were calculated for the NFI data, and the results were then compared to those of the TASO data. In the calculations of the potential allowable cut from NIPF lands, it was assumed that timber management strategies based on NFI data would represent the area-based proportion of the choice of each strategy in the TASO data. The main characteristics of the sample are presented in Table 2.

Growth and removals of both TASO and NFI data were made current to the beginning of year 1994. Corrections were done using the MELA system. Without correcting the data sets, comparison of information on forest resources and timber management strategies would have been difficult because the TASO data originated from the years 1989-1993 and the NFI data from 1990. In addition, correcting the data enabled the use of data that are as recent as possible. Constraints used in the optimisations were annual removals based on statistics by timber assortment and the harvest areas by harvesting method.

After updating growth and removals, the initial volume of the growing stock, (an average of the sample holdings) was 120.6 m^3/ha and in the NFI data 122.0 m^3/ha (Table 3). The mean initial volumes of both the TASO and NFI data sets were very close to each other. The TASO data included more pine and spruce, but less birch than the NFI data. Furthermore, there was more sawtimber in the NFI data than in the TASO data.

4. POTENTIAL ALLOWABLE CUT IN FINLAND

In the maximisation of the utility function (2.1), the most preferred strategy obtained was "Sustainability" (chosen by 61% of landowners). The second-most preferred was "Finance" (17%) and the third was "Saving" (14%). "No cuttings" and "Max cuttings" were the least preferred (4% each). When presented according to the number of landowners, the distributions of the most preferred strategies were slightly different than when compared to the forest area represented by each strategy (Figure 3).

Table 2. Main characteristics of the sample, based on owners' responses.

	Mean	SD
OWNER, %		
Farmer	51.4	
Non-farmer	48.6	
AGE, a	50.9	13.0
FOREST AREA, ha	57.2	58.7
ARABLE LAND	14.6	23.2
PRODUCTION ORIENTATION, %		
Agriculture	11.3	
Agriculture and forestry	41.7	
Forestry	34.5	
Recreation and residence	12.5	
TIMBER PRODUCTION POSSIBILITIES, %		
Good	43.2	
Fairly good	43.4	
Poor	13.4	
FUTURE CUTTINGS, %		
Extensive cuttings	32.0	
Sustainability	55.3	
Intensive cuttings	12.7	

Table 3. Volume of growing stock (m^3/ha) by tree species in the TASO and NFI data.

	TASO	NFI8
Average	120.6	122.0
Scots pine	47.8	45.4
Norway spruce	58.3	54.4
Hardwood	14.5	22.2
Sawtimber	52.6	54.0

The timber management strategies were compared at the regional level assuming that all landowners would follow the same strategy. Comparisons were made for both the TASO and the NFI data sets to verify the reliability of the TASO data in southern Finland. Average removals in both data sets were compared over the entire 20-year planning horizon.

In the "Sustainability" strategy, the average harvest rate in the NFI data was 1.2% higher compared to the TASO data (Table 4a). In the "Saving" strategy, the average harvest was, by definition, approximately half of the removals of the "Sustainability" strategy. In the "Finance" strategy, the average removals were smaller than in the "Sustainability" strategy and in the case of the "Finance" strategy, the removals in the TASO data were 6.9% greater than those in the NFI data.

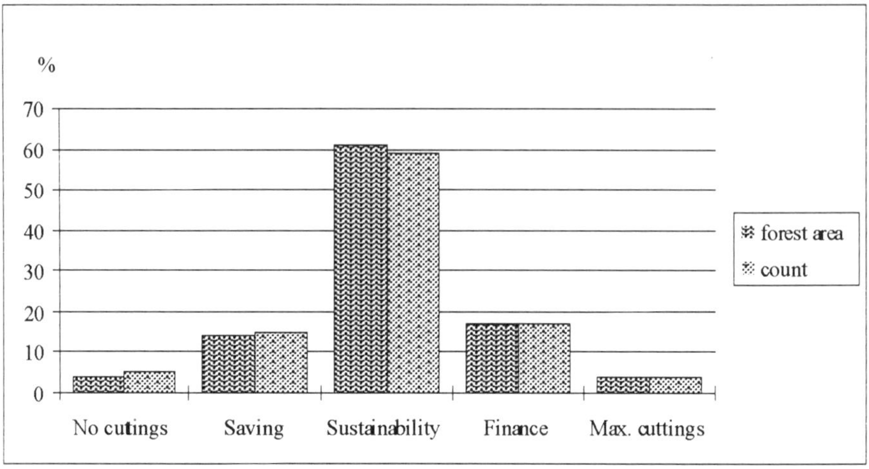

Figure 3. Choices of timber management strategies by number of landowners, and actual forest area represented by each strategy.

Table 4. Average removals over 20 years, according to strategy, m^3/ha (a), and cumulative removals, mill. m^3, in southern (b) and northern (c) Finland, assuming that all landowners would follow the same strategy

(a) Average removals (m^3/ha) in southern Finland

	TASO	NFI
No cuttings	0.00	0.00
Saving	2.51	2.51
Sustainability	4.96	5.02
Finance	4.66	4.36
Max Cuttings	5.11	4.36

(b) Cumulative removals (mill. m^3) in southern Finland

	TASO	NFI
No cuttings	0.00	0.00
Saving	22.40	22.47
Sustainability	44.36	44.88
Finance	41.68	38.93
Max cuttings	45.70	38.97

c) Cumulative removals (mill. m^3) in northern Finland

	TASO	NFI
No cuttings	-	0.00
Saving	-	4.21
Sustainability	-	8.41
Finance	-	7.37

When assuming that all landowners would choose the "Max cuttings" strategy, the average removals were considerably greater in the TASO data than in the NFI data. However, the difference had only a small effect on the potential allowable cut, since only 4.0% of the landowners had chosen the "Max cuttings" strategy. The greatest cumulative removals were obtained in the TASO data with the "Max cuttings" choice, and in the NFI data when choosing "Sustainability" (Table 4 b). The results for northern Finland were presented only with NFI data, because the mail questionnaire was conducted in southern Finland (Table 4 c).

To generalise the results for the whole of Finland, the potential allowable cut of the 20-year planning horizon was 46.5 mill. m^3/yr for the TASO data and 46.3 mill. m^3/yr for the NFI data (Figure 4). During the first half of the planning period, the removals were heavier due to the accumulation of removals in the "Max cuttings" and "Finance" strategies. The proportion of sawtimber in the removals was somewhat higher in the TASO data than in the NFI data.

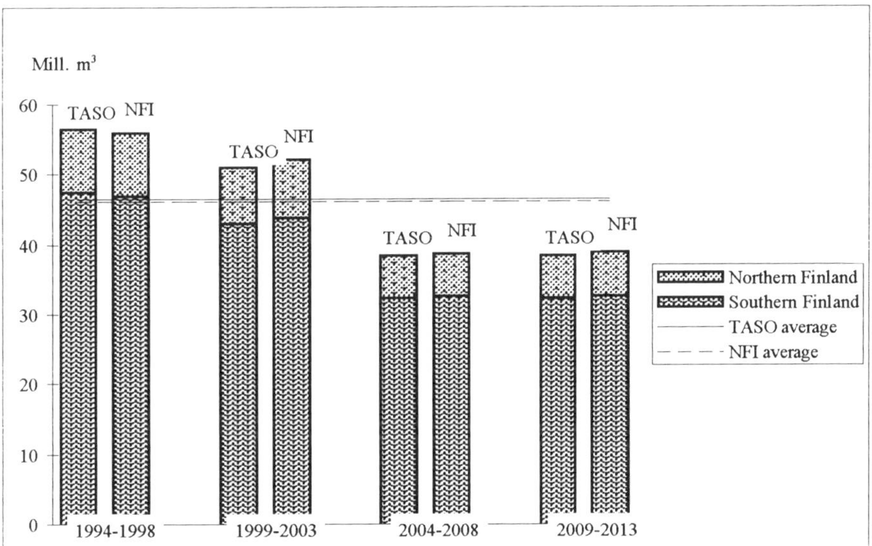

Figure 4. Removals according to choices of preferred strategies during each five-year period and average removals.

In both data sets, the mean volume increased toward the end of the planning horizon (Figure 5). In addition, the mean volume in the TASO data was a little higher, particularly in the proportion of sawtimber. This was due to the higher level of growth observed in the TASO data when compared to that in the NFI data (Figure 6).

5. DISCUSSION AND CONCLUSIONS

5.1 Comparing the Cutting Budgets

The potential allowable cut determined in this study was 18.9% higher than average actual harvests during the years 1989-1994 (Figure 7). However, during the recent economic boom, the potential allowable cut is at the same level of removals. Compared to the greatest allowable cut (based on sustained yield) of NFI8, the potential allowable cut of this study was 12.7% smaller. Furthermore, the cutting budget based on combining the forestry plans was 12% smaller than the one presented in this study.

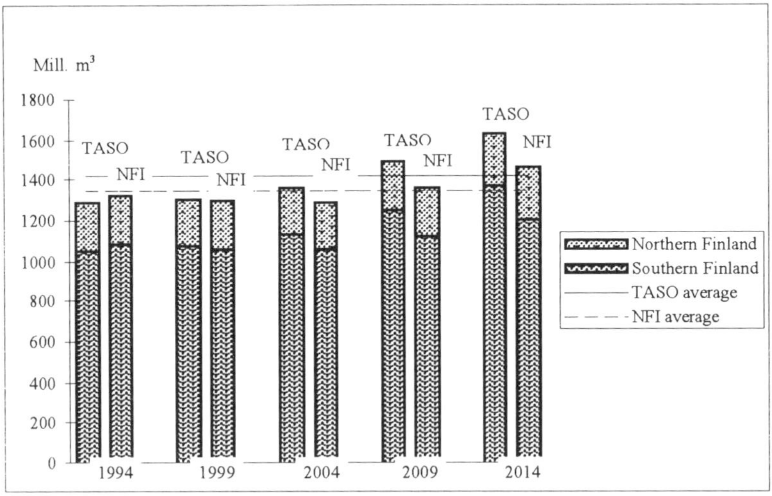

Figure 5. Total volume according to planning periods, and average volume over a period of 20 years.

Differences between cutting budgets based on combining forestry plans and potential allowable cut as defined in this study are due to two reasons: (1) the principle of discretion in NIPF planning and (2) the older, NFI7-based growth models used in the TASO planning system. Underestimation of the actual cutting possibilities on a sustained-yield basis in the TASO forestry plans can be almost 20% (Pesonen and Räsänen 1993). Forest planning of NIPFs is still based on stand-wise suggestions for treatments made by professional planners. Planners seldom have full knowledge of the sustained cutting possibilities at the forest-holding level.

In comparing the NFI8-based, forest-resources-oriented cutting budget and the potential allowable cut of this study, two main reasons for the difference can be offered: (1) ignorance of landowner-specific forestry goals in the former and (2) constraints caused by the requirement of forest-holding level sustainability in the latter. The fact that the owners of small forest holdings preferred the choices of "No cuttings" and "Saving" strategies reduces the potential allowable cut from NIPF lands. The requirement of sustained yield at the forest holding level has been reported to decrease regional cutting levels by over 10% (Pesonen and Soimasuo 1998).

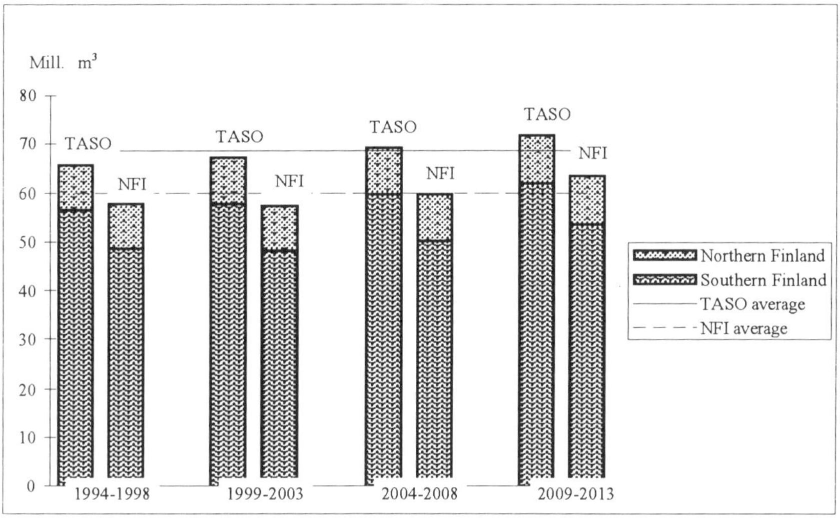

Figure 6. Increment according to time periods and average increment.

5.2 The Reliability of the Data and Methods

The additive utility function, the function form used in this study, is the easiest to interpret (Pukkala and Kangas 1993). It has been noticed in several studies that the additive utility function produces a utility index which best describes the preferences of the decision maker (Tell 1976, Laskey and Fischer 1987). It has also been stated that landowners are utility maximisers, who consider both the economic and the non-economic benefits of their forests (Boyd 1984, Hyberg 1987).

Due to its simplicity, effectiveness, and ability to deal with qualitative as well as quantitative criteria (this is also indicated by the results of this study), the AHP is well-suited to dealing with problems in forest management planning (e.g. Kangas 1992). When used in mail questionnaires, its main weakness lies in the question of whether all respondents are able to concentrate on the numerous comparisons required by the AHP. Therefore, the results might be improved by the application of personal interviews in conjunction with data collection.

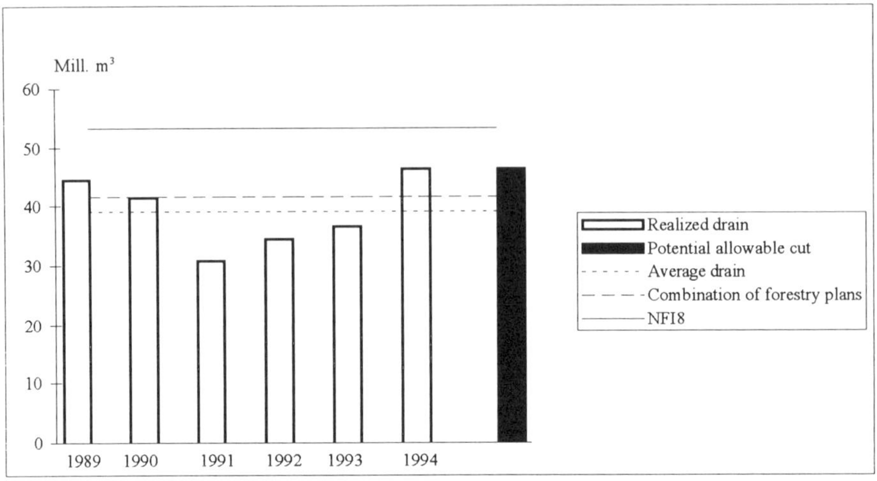

Figure 7. Realised drain, and cutting budgets calculated using alternative methods.

The consistency ratios (CR) were slightly higher than was acceptable: 16.8% in economic and 15.2% in non-economic comparisons. This may be partly due to the fact that the questionnaires used in the study were mailed, and variously skilled and committed landowners did dozens of AHP comparisons. However, there is no unequivocal upper limit for the level of inconsistency in pairwise comparisons, and moreover, the inconsistency in comparisons can also be due to a conscious choice, and therefore can be accepted (e.g., Wedley 1993, Apostolou and Hassell 1993).

The AHP method and the use of a mail questionnaire in data collection limited the alternative choices of strategies to five. In spite of this, the alternative strategies and choices made by the landowners were based on the actual, forest-holding-level development of cuttings, income from timber sales, and other forest characteristics. Although few landowners chose the extreme alternatives–"No cuttings" or "Max cuttings" strategies–these strategies were included in the comparisons in order to describe the whole range of timber production possibilities.

Objectives of the landowners may also vary temporally and geographically. Lönnstedt and Törnqvist (1990) stated that the choice of timber management strategy is affected by the needs and objectives of both short- and long-term perspectives. The goal structure of the landowners could have been clarified better. In this study, the landowners were able to compare only five precalculated timber management strategies. It would be possible to ask more specific questions about the objectives of the landowners in the first phase of the questionnaire, and with that information in mind, calculate the strategies more individually.

Overall differences between the results from the two data sets (TASO and NFI) were small. Based on this study, data from the standwise inventory is reliable enough to enable the definition of regional cutting possibilities, although considerable measurement errors have been reported in stand-wise inventories due to the personal (subjective) characteristics of the planners (Laasasenaho and Päivinen 1986). TASO data appeared representative in comparison to the reference material (the NFI data) with respect to the forest resource information. No major differences between the data sets were found for mean volumes, proportions of sawtimber and tree species, and age-class distributions. The only substantial difference was greater growth present in the TASO data, which was partly due to the greater proportion of seedling stands compared to the NFI data. In "Sustainability," for example, the growth in the TASO data was 14.8% greater during the whole planning horizon than in the NFI data.

Due to sustainability requirements at the forest-holding level, removals resulting from the TASO data were smaller and led to faster volume increases and higher growth rates. One reason for the difference could have been that, in the NFI data, diameter distributions were constructed empirically, while in the TASO data, diameter distributions were formulated using the theoretical, Weibull distribution (Kilkki *et al.* 1989). The reliability of the results could have been further improved by selecting diameter distributions from the NFI data by using the standwise information of the TASO data.

5.3 Conclusions

The potential allowable cut presented in this study appears to fall between the actual harvest and the greatest allowable cut based on the National Forest Inventory. Results indicate that the landowners' future harvesting intentions will ensure an availability of wood material for forest companies. Furthermore, the region's landowners could be encouraged to practice intensive management and harvesting by demonstrating strategic alternatives for timber management. The results of this study may also help to direct the development of management planning on NIPF lands.

An interesting issue for future research would be to monitor the sample forest holdings: do strategic calculations affect future harvesting behaviour of the owners? In addition, forestry plans based on the choices of timber management strategies could be made for the sample holdings and then proceed to monitor owner's harvesting behaviour.

REFERENCES

APOSTOLOU, B. AND HASSELL, J. M. 1993. An empirical examination of the sensitivity of the analytic hierarchy process to departures from recommended consistency ratios. Mathematical and Computer Modelling 17: 163-170.

BOYD, R. G. 1984. Government support of nonindustrial production: The case of private forests. South. J. Econ. 51:89-107.

FOREST 2000 PROGRAM. 1985. Main Report. Council of Economics. Section of the Forest 2000 Program.

HANSSON, K., LÖNNSTEDT, L. AND SVENSSON., J. 1990. Decision support system for nonindustrial private forest owners. Swedish University of Agricultural Sciences. Dep. For.Ind.Market Studies. Rep. 12.

HIRSHLEIFER, J. 1984. Price theory and applications. Prentice-Hall. 536 p.

HYBERG, B. T. 1987. Multiattribute decision theory and forest management: A discussion and application. Forest Science 33: 835-845.

JOHNSON, K. N. AND JONES, D. B. 1986. A user's guide to multiple use - sustained yield resource scheduling calculation (MUSYC): Timber management, USDA Forest Service, Washington DC.

JONSSON, B., JACOBSSON, J. AND KALLUR, H. 1993. The forest management planning package. Theory and application. Studia Forestalia Suecica 189. 56 p.

KAJANUS, M, KURTTILA, M. AND PESONEN, M. 1998. A simultaneous equation model for predicting the production frontier of a forest holding. Canadian Journal of Forest Research 28: 1851-1860.

KANGAS, J. 1992. Multiple-use planning of forest resources by using the Analytic Hierarchy Process. Scand. J. For. Res. 7: 259-268.

KANGAS, J. AND PUKKALA, T. 1992. A decision theoretic approach applied to goal programming of forest management. Silva Fennica 26:169-176.

KAST, E. K. AND ROSENZWEIG, J. E. 1974. Organization and management. A systems approach. Second edition. 389 p.

KILKKI, P. 1968. Income-oriented cutting budget. Acta Forestalia Fennica 91:1-54.

KILKKI, P., MALTAMO, M., MYKKÄNEN, R. AND PÄIVINEN, R. 1989. Use of the Weibull function in estimating the basal area dbh-distribution. Silva Fennica 23: 311-318.

KILKKI, P. AND SIITONEN, M. 1976. Principles of a Forest Information System. XVI IUFRO World Congress, Division IV. Proceedings. p. 154-163.

KURTTILA, M, PESONEN, M., KANGAS, J. AND KAJANUS, M. 1998. Utilizing the analytical hierarchy process (AHP) in SWOT analysis - A hybrid method and its application to a forest-certification case. Forest Policy and Economics. In print.

LAASASENAHO, J. AND PÄIVINEN, R. 1986. Kuvioittaisen arvioinnin tarkistamisesta. Summary: On the checking of inventory by compartments. Folia Forestalia 664. 19 p.

LASKEY, K. B. AND FISCHER, G. W. 1987. Estimating utility functions in the presence of response error. Manage. Sci. 33:965-980.

LÖNNSTEDT, L. 1989. Goals and cutting decisions of private small forest owners. Scand. J. For. Res. 4:259-265.

LÖNNSTEDT, L. AND TÖRNQVIST, T. 1990. The Owner, the Estate and the External Influences: Nonindustrial private forest owners' management decisions. Swedish University of Agricultural Sciences. Dep. of Forest-Industry-Market Studies. Report No. 14. Uppsala, Sweden.

LÖNNSTEDT, L AND ROOS, A. 1993. Cutting Levels for Nonindustrial Forestry: An estimate based on the state of forests and the goals of forest owners. Swedish University of

Agricultural Sciences. Dep. of Forest-Industry-Market Studies. Report No. 26. Uppsala, Sweden.

MENDOZA, G. A. AND SPROUSE, W. 1989. Forest planning and decision making under fuzzy environments: An overview and illustration. Forest Science 35: 481-502.

MYKKÄNEN, R. 1994. Aspiration-based utility functions in a planning model for timber flow manangement. Acta Forestalia Fennica 245. 66 p.

PESONEN, M. 1995. Nonindustrial private forest landowners' choices of timber management strategies and potential allowable cut. Acta Forestalia Fennica 247. 31 p.

PESONEN, M. 1996. Estimation of potential allowable cut using modeling of landowners' strategic decision making. Dissertation. Finnish Forest Research institute, Research Papers 625. 56 + 83 p.

PESONEN, M., KETTUNEN, A., HEIKKINEN V.-P. AND RÄSÄNEN, P. 1994. Yksityismetsänomistajien puuntuotantostrategiat ja potentiaaliset hakkuumahdollisuudet Pohjois-Savossa. Metsäntutkimuslaitoksen tiedonantoja 515. 50 p.

PESONEN, M. AND RÄSÄNEN, P. 1993. Metsäverovalinta - strateginen ratkaisu. Metsäntutkimuslaitoksen tiedonantoja 472. 78 p.

PESONEN, M. AND SOIMASUO J. 1998. Tilakohtaisen kestävyyden vaikutus suuralueen kestäviin hakkuumahdollisuuksiin - tapaustutkimus Satakunnan metsälautakunnan alueella. Metsätieteen aikakauskirja - Folia Forestalia 1/1998: 43-51.

PIHLJERTA, K. 1994. Maatilalaskennan tilastoja 1988-1994. Verohallitus.

PUKKALA, T. AND KANGAS, J. 1993. A heuristic optimization method for forest planning and decision making. Scandinavian Journal of Forest Research 8: 560-570.

RANTA, R. 1991. Metsätalouden suunnittelulaskelmat. In: Tapion taskukirja. 21. painos. Kustannusosakeyhtiö Metsälehti. p. 334-337.

SAATY, T. L. 1977. A scaling method for priorities in hierarchical structures. Journal of Mathematical Psychology 15: 234-281.

SAATY, T. L. 1980. The Analytic Hierarchy Process. McGraw-Hill. New York. 287 p.

SEVOLA, Y. (EDS.) 1998. Metsätilastollinen vuosikirja 1998-Skogsstatistisk årsbok-Finnish Statistical Yearbook of Forestry. 348 p. SVT Maa- ja metsätalous 1997:4. 348 p.

SIITONEN, M. 1983. A Long-Term Forestry Planning System Based on Data from the Finnish National Forest Inventory. Proceedings of the IUFRO subject group 4.02 meeting in Finland. University of Helsinki, Dep. of Forest Mensuration and Management, Research Notes 17. p. 195-207.

SIITONEN, M. 1993. Experiences in the Use of Forest Management Planning Models. Silva Fennica 27 (2). p. 167-178.

TELL, B. 1976. A comparative study of some multiple-criteria methods. The Economic Research Institute at the Stockholm School of Economics. 203 p.

WARDLE, P. A. 1965. Forest management and operational research: a linear programming study. Ser. B, Management science 11: 260-270.

WARE, G. O. AND CLUTTER, J. L. 1971. Mathematical programming system for the management of industrial forests. Forest Science 17(4): 428-445.

WEDLEY, W. C. 1993. Consistency prediction for incomplete AHP matrices. Mathematical and Computer Modelling. 17: 151-161.

VON WINTERFELDT, D. AND EDWARDS, W. 1986. Decision analysis and behavioral research. Cambridge University Press. Cambridge. 604 p.

Chapter 12

Applying A'WOT to Forest Industry Investment Strategies: Case Study of a Finnish Company in North America

Mauno Pesonen
Finnish Forest Research Institute, Finland

Jyrki Ahola
Lappeenranta University of Technology, Department of Business Administration, Finland

Mikko Kurttila
University of Joensuu, Faculty of Forestry, Finland

Miika Kajanus
Pohjois-Savo Polytechnic, Rural Education, Peltosalmi, Finland

Jyrki Kangas
Finnish Forest Research Institute, Kannus Research Station, Finland

Key words: A'WOT, decision analysis, external and internal environment, forest industry, investment, strategic planning, SWOT

Abstract: The present study examines a hybrid method, referred to here as the hybrid method A'WOT, for improving the usability of SWOT (strengths, weaknesses, opportunities, and threats) analysis. A commonly–used decision analysis method, the analytic hierarchy process (AHP), and its eigenvalue calculation framework are integrated with SWOT analysis. AHP's connection to SWOT yields analytically-determined priorities for the factors included in SWOT analysis and makes them commensurable. The aim in applying the hybrid method is to improve the quantitative information basis of strategic planning processes. The hybrid method was tested in connection with a Finnish forest

D.L. Schmoldt et al. (eds.),
The Analytic Hierarchy Process in Natural Resource and Environmental Decision Making, 187–198.

industry enterprise's decision to invest in North America. In the case study, the results were presented in an illustrative way by utilising the quantitative information achieved by the hybrid method. The results indicated that forest-industry investment was reasonable in North America. In addition, the required pairwise comparisons were found to be useful, because they force the decision-maker to think over the weights of the factors and to analyse the situation more precisely and in more depth.

1. INTRODUCTION

1.1 Strategies of the Finnish Forest Industry

Forest industry enterprises, like other businesses, are influenced by changes within the internal (e.g. change in productivity) and external (e.g. change of government) operational environments. Common strategic planning approaches are fundamentally based on adjusting to changes in the external environment and there exists a wide range of planning methods that are well suited for simultaneous analysis of the interactions of both environments. The enterprise's strategy process is seen as a way to consider, decide on and implement strategies (Ahola 1995). The strategy process does not form a sequential, hierarchical system, but rather a group of activities to be implemented when the need arises. The strategy process consists of management's working process to produce such strategies as will fulfil the owner's and other major stakeholders' objectives for the enterprise.

The life cycles of forest industry products are very long. Also, the business cycles of forest industry enterprises are long. Nevertheless, this branch of industry is very sensitive to business trends and the demand for these products can change quickly. Considering these features, the forest-industries sector is not the easiest of business sectors to research. Moreover, this is also a very capital intensive branch of industry, because the production units tend to grow bigger and bigger. Starting up of pulp and paper machines creates new production capacity, with many years having gone into planning and construction. Therefore, the correct timing of investment decisions is of crucial importance. The starting point of investment can be explained by evolutionary process theory (Ahola 1995) which explains changes in organisations and society. It does not, however, predict changes; it only explains them.

Globalisation and the increasing size of forest industry companies has made investment-related decision making more difficult because of financial (Finchem 1997) and cultural issues (Very *et al.* 1997, Herbert 1999) involved. Despite this, business areas have been fragmented into smaller groups according to business, localisation, and other criteria. The goal has been to bring more flexibility into production, to accelerate decision making, to seek out synergies, and to achieve better control over competition: success in competition requires concentration on key knowledge and products. Forest industry production is moving closer to the customers (Rinne 1995, Geginat 1998). The vision of Finnish forest industries can be summarised as follows: 1) good profitability and balance, 2) customer orientation, 3) structural changes, 4) concentration of business, and 5) globalisation.

Growth in production has taken place as the result of acquisitions, joint equity ventures, joint contractual ventures, and organic growth. Acquisitions are an easy way to make an entry into new markets while the goal of joint equity ventures is to win new markets by minimising risks. An example of an acquisition was the case of UPM-Kymmene of Finland acquiring the North American Blandin Paper Company. The seller was Fletcher Challenge Canada Limited, a member of the Fletcher Challenge Group of New Zealand. The acquisition price was USD 650 million. The implementation of a fine paper alliance, between UPM-Kymmene and APRIL, a Singapore-based Asian group, was an example of a joint equity venture. The pulpmill of Metsä-Rauma Ltd, jointly owned by Metsä-Serla and UPM-Kymmene, is an example of a joint contractual venture.

International trade has grown rapidly and increasingly intensifying competition has led to production being relocated to places closer to markets. Internationalisation has also contributed to companies having a better capacity to serve customers, shrinking distances, faster deliveries, and more recycling of paper (Higham 1996). Mainly as a result of acquisitions, Finnish forest-based manufacturers now own 43 paper and board mills in Western Europe, four in North America, one in South America and two in Asia. Fifty percent of the production capacity owned by Finnish forest-based companies was located abroad in 1998. The companies' sales organisations spanned the globe. This shows that seeking new market areas (e.g. in North America and Asia) is a logical way to globalise and increase the size of one's business.

1.2 SWOT and AHP

SWOT (an acronym standing for strengths, weaknesses, opportunities and threats) analysis is a commonly used tool for analysing internal and external environments in order to attain a systematic approach and support for a decision situation. Some examples of weighting SWOT factors have also been presented (e.g. Kotler 1988, Wheelen and Hunger 1995). SWOT analyses in particular have their mutual origins in the work of business policy academics at the Harvard Business School and other American business schools from the 1960's onward. The work of Kenneth Andrews

(Andrews 1971, 1980) has been especially influential in popularising the idea that good strategy means ensuring a fit between the external situation that a firm faces (threats and opportunities) and its own internal qualities or characteristics (strengths and weaknesses).

The internal and external factors most important to the enterprise's future are referred to as strategic factors and these are summarised within SWOT analysis. The final goal of the strategic planning process, of which SWOT is an early stage, is to develop and adopt a strategy resulting in a good fit between internal and external factors. SWOT can also be used when a strategy alternative emerges suddenly and the decision context relevant to it has to be analysed.

If used correctly, SWOT can provide a good basis for successful strategy formulation. Nevertheless, it could be used more efficiently (McDonald 1993). When using SWOT, the analysis lacks the possibility of comprehensively appraising the strategic decision-making situation; it merely pinpoints the number of factors in strength, weakness, opportunity or threat groups but does not pinpoint the most significant group. In addition, SWOT includes no means to analytically determine the importance of factors or to assess the fit between SWOT factors and decision alternatives. The further utilisation of SWOT is, thus, mainly based on the qualitative analysis, capabilities and expertise of the persons participating in the planning process. As numerous criteria and interdependencies often complicate planning processes, it may be that the utilisation of SWOT is insufficient. In their study, Hill and Westbrook (1997) found that none of the twenty companies prioritised individual SWOT factors; one grouped factors further into subcategories, and only three companies used SWOT analysis as an input for a new mission statement. In addition, the expression of individual factors was of a very general and brief nature. Thus, it can be concluded that the result of SWOT analysis is too often only a superficial and imprecise listing or an incomplete qualitative examination of internal and external factors.

Kurttila *et al.* (1999) examined a new hybrid method for improving the usability of SWOT analysis; we refer to it in this paper by the acronym A'WOT. A commonly used decision analysis method, the analytic hierarchy process (AHP), and its eigenvalue calculation framework were integrated with SWOT analysis. AHP's connection to SWOT yields analytically determined priorities for the factors included in SWOT analysis and makes them commensurable.

The present study deals with A'WOT to analyse a Finnish forest industry company's investment decisions in North America. As a result, a clearer picture, including quantitative information, of the factors affecting investment decisions of forest industry is created.

2. A'WOT (AHP IN SWOT ANALYSIS)

Basically, the results of an AHP analysis are the overall (global) priorities of decision alternatives. The idea in utilising AHP within a SWOT framework is to systematically evaluate SWOT factors and commensurate their intensities, which can be regarded as valuable additions to SWOT analysis. Additional value from a SWOT analysis can be achieved by performing pairwise comparisons between SWOT factors and analysing them by means of the eigenvalue technique as applied in AHP. This offers a good basis for examining the present or anticipated situation, or a new strategy alternative, more comprehensively.

To help in understanding the A'WOT method, the following definitions are used: SWOT *groups* refer to the four entities (i.e. strengths, weaknesses, opportunities and threats) and SWOT *factors* refer to the individual factors underlying these groups. The A'WOT method proceeds as follows:

Step 1. SWOT analysis is carried out. The relevant factors of the external and internal environment are identified and included in SWOT analysis. When the relative ranking technique of the AHP is applied, it is recommended that the number of factors within a SWOT group should not exceed 10 because the number of pairwise comparisons needed in the analysis increases rapidly (Saaty 1980). See chapter 2 of this volume for more details regarding the scoring techniques available in the AHP.

Step 2. Pairwise comparisons between SWOT factors are carried out within each SWOT group. When making the comparisons, the questions at stake are (1) that of the two factors compared has greater importance and (2) how much greater. Using these comparisons, the relative local priorities (importance) of the factors are computed using the eigenvalue method.

Step 3. Pairwise comparisons are made between the four SWOT groups. The factor with the highest local priority is chosen from each group to represent the group. These four factors are then compared and their relative priorities are calculated as in Step 2. These are the scaling factors of the four SWOT groups and are used to calculate the overall (global) priorities of the independent factors within groups. This is done by multiplying the factors' local priorities (defined in Step 2) by the value of the corresponding scaling factor of the SWOT group. The global priorities of all the factors sum up to one.

Step 4. The results are utilised in the strategy formulation and evaluation process. The contribution to the strategic planning process comes in the form of numerical values for the factors. New goals may be set, strategies defined, and implementations planned to take into consideration the prominent factors.

3. RESULTS OF THE CASE STUDY

Construction of the case-study SWOT framework was carried out by collecting information from numerous forest-industry-related publications (e.g. Higham 1996, Payne 1997). Experts in the corporation were interviewed (one director in business development and two senior vice-presidents) to get specific information and to strengthen our preconception of the company's US-investment strategies. In this analysis, pair-wise comparisons were carried out by one person, whose expertise covers the strategic planning process and development of the companies' international investment decisions (Figure 1).

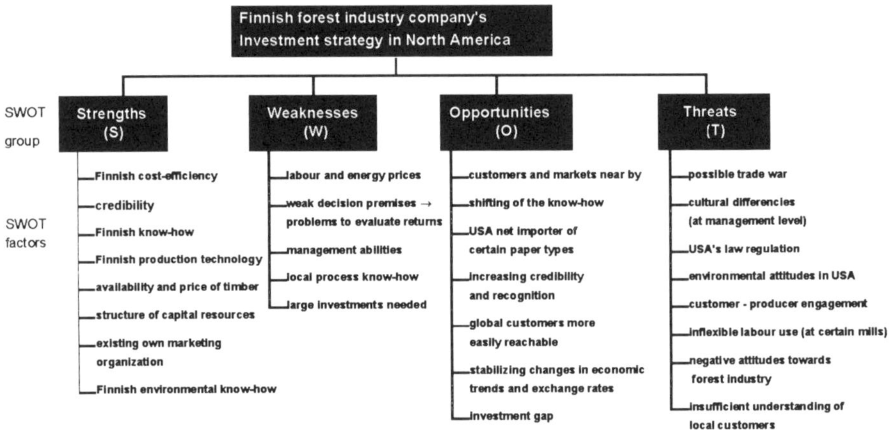

Figure 1. Decision hierarchy of an investment strategy of Finnish forest industry.

Opportunities that are seen to exist in North America and the strengths that are considered to be utilised in the market area are logically predominating (Figure 2). Customer- and market-oriented strategies highlight the vicinity of existing and potential customers and markets (Table 1). Entering a new market area includes risks, and the available information is always insufficient and therefore evaluations of expected returns are exceptionally uncertain, which, in this case, is the foremost weakness. However, it is expected that becoming established in the market area will, in the future, serve to significantly decrease some threats, e.g. misunderstanding of local customers. The consistency ratios were greater than 0.10 in all but one case.

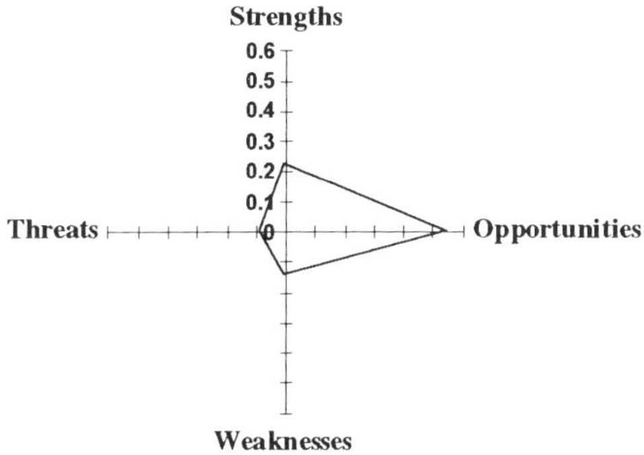

Figure 2. The relationships of SWOT groups in the forest industry case study determined by the priorities of the scaling factors.

4. DISCUSSION

4.1 Case Study Conclusions

According to the results of SWOT analysis and pairwise comparisons, strengths and opportunities dominate the operational environment affecting North American investments by this Finnish forest industry company. Proximity to customers and markets are the most important opportunities, and Finnish production technology is the greatest strength for becoming established in this market area.

Establishing itself in North America is a logical decision by Finnish forest industries as an element of globalisation. If Finnish forest industry enterprises want to be part of this process, they have to get themselves into the North American markets, albeit that the United States and Canada are not countries promising "great adventures" in the sense that Asia possess great potential returns and great risk.

The main goal in establishing business in North America is to comply with the client-orientation of markets. Consumers demand local and quick service. Both local presence and production credibility strengthen the competition advantage of a foreign company. Moreover, potential trade wars and customs barriers do not threaten local companies. Also, North America is a net exporter of value-added paper products only and even then only to a small degree. The Nordic countries' cost-efficiency, know-how, and production technology are strengths when operating in the North American markets.

Table 1. Priorities and consistency ratios[1] of comparisons of the SWOT groups and factors (the factor having the greatest weight in each SWOT group is underlined). The overall priority of the factor is computed by multiplying the priority of the factor within the group by the priority of the group.

SWOT group	Group priority	SWOT factors	Consistency ratio	Priority of factor within group	Overall priority of factor
Strengths	0.223	Finnish cost efficiency		0.134	0.030
		Credibility		0.034	0.008
		Finnish know-how		0.218	0.049
		Finnish production technology	0.080	0.231	0.052
		Availability and price of timber		0.072	0.016
		Structure of capital resources		0.042	0.009
		Existing own marketing organisation		0.203	0.045
		Finnish environmental know-how		0.063	0.014
Weaknesses	0.143	Labour and energy prices		0.045	0.006
		Weak decision premises -> problems in evaluating returns	0.102	0.325	0.046
		Management skills		0.378	0.054
		Local process know-how		0.070	0.010
		Large investments needed		0.181	0.026
Opportunities	0.545	Customers and markets close by		0.278	0.152
		Shifting of know-how		0.082	0.045
		USA net importer of certain paper grades		0.143	0.078
		Increasing credibility and recognition	0.173	0.176	0.096
		Global customers more readily reachable		0.111	0.060
		Stabilising changes in economic trends and exchange rates		0.097	0.053
		Investment gap		0.112	0.061
Threats	0.088	Possible trade war		0.044	0.004
		Cultural differences (at management level)		0.090	0.008
		USA's legislation		0.054	0.005
		Environmental attitudes in USA	0.103	0.114	0.010
		Customer-producer engagement		0.268	0.024
		Inflexible labour use (at certain mills)		0.110	0.010
		Negative attitudes towards forest industries		0.071	0.006
		Insufficient understanding of local customers		0.247	0.022

[1] The consistency ratio of the comparisons between four SWOT groups was 0.123.

Consistency ratios were high and comparisons were difficult to make because of the large number of SWOT factors and their sometimes difficult interpretations. Furthermore, the use of a *representative*, group factor can obfuscate group comparisons. Following comparisons of factors within each group, one way could be to compare these groups directly to each other.

4.2 Evaluation of A'WOT

The present study tried to explain the investment strategies of a Finnish forest industry enterprise in North America using SWOT and the analysis was deepened by producing quantitative information of their importance by utilising the characteristics of the AHP. Although SWOT is in common use as a planning tool, it has some weaknesses. A hybrid application called A'WOT was used to mitigate some of the weaknesses of SWOT.

Due to its simplicity, effectiveness and ability to deal with qualitative as well as quantitative criteria (this was also indicated by the results of this study), the AHP is well suited to deal with the factors in SWOT analysis. One problem with SWOT analysis is uncertainty of future events and the outcome of different factors. This complicates present-day comparisons. However, AHP analysis is capable of handling decision-making situations with some uncertainties and inconsistencies.

Using relative ranking in the AHP, the number of factors within the strengths, weaknesses, opportunities or threats should be limited to ten, as this probably induces the user to avoid overlapping and carelessness when constructing SWOT lists. On the other hand, the limitation is not so strict, and the problem of having a large number of comparisons can be avoided by several different techniques. First, alternative scoring methods in the AHP (e.g., absolute ratings and benchmark ranking) can be used. Second, variables can be grouped, which adds a new level to the comparison hierarchy (Saaty 1980). If, for example, the number of opportunities is large, they can be grouped into two or three subgroups. Opportunities, for example, may be divided into "General Environmental Opportunities" and "Competitive Environmental Opportunities" (Dess and Miller 1993). Third, new data recording and analysis techniques offer possibilities to include more factors in decision analysis. For example, Alho and Kangas (1997) presented a regression version of the AHP formulated in Bayesian terms. Their approach can be developed and utilised so that not all comparisons need to be performed. See chapter 15 of this volume for additional information.

The AHP provides quantitative priorities to be used in decision support. It does not, however, include statistical assessment of uncertainties of the results. The measure of the consistency of the comparisons made, the consistency ratio, resulting from AHP calculations provides no direct information about the uncertainty of the priorities obtained. Other methods for analysing uncertainties in pairwise comparisons have been presented. Alho *et al.* (1996) suggested a variance-components modelling approach, where uncertainty or variation of the judgements in the case of multiple judges can be divided into three parts: (1) interpersonal variation around the population mean; (2) possible shared logical inconsistency of the judgements among the judges; and (3) residual uncertainty. Alho and Kangas (1997) extended this formulation to a multilevel, multiple-objective choice problem by using regression technique and the Bayesian approach. As a result, it is possible to attach probability to the resulting priorities. These techniques might also be used in A'WOT.

Numerical priorities of SWOT factors, are useful when formulating or choosing a strategy. It is useful to compare the external possibilities in relation to the internal capabilities, because all factors are on a commensurable numerical scale. For example, when it is observed that one single weakness is bigger than all the strengths, the strategy chosen could perhaps be aimed at eliminating this weakness. Similarly, choosing a new strategy should probably not be based merely on opportunities while omitting the existing threats of equal magnitude.

Results of our case study were presented in an illustrative way, which is often needed to clarify the interactions of numerous and contradictory factors. In strategic planning, this is often implemented by means of matrixes or graphs. Well-known examples of these instruments are the Boston Consulting Group's *Business Portfolio Matrix* (business growth rate and relative competition position), General Electric's approach (market attractiveness and competitive position), and Ansoff's *product/market expansion grid*, and others (e.g. Ansoff 1965, Hofer and Schendel 1978, Dess and Miller 1993).

The hybrid method presented here is suitable for many kinds of strategic planning situations. Kurttila *et al.* (1999) used this hybrid method in connection with a Finnish case study on forest certification. Results indicated that certification could be a potential strategic alternative for Finnish farms with adjoining forestry. The method was also used in natural resource planning by the Finnish Forest Park Service in western Finland (Kurttila *et al.* 1998).

In this case study, the situation investigated was one where a new strategy option emerged. The method can also be used in situations where strategic options have not yet been created. After creating priorities for the SWOT factors, new strategies can be constructed based partly on priority information. It might also be possible to incorporate Weihrich's (1982) and Proctor's (1992) techniques utilising priorities to determine the most important factors for creating new strategies.

One approach to dealing with the uncertainties involved in the assessment of future development might be the application of scenario modelling. In this approach, each possible future scenario would have its own SWOT analysis and AHP comparisons. Appraising the probabilities to scenarios and weighting the SWOT factors with them could yield a more comprehensive picture of the effects of the various future outcomes. Weihrich (1982), too, proposed a dynamic SWOT analysis, where changes in internal and external factors are compared over time.

Based on the experiences of this study, the combined use of the AHP and SWOT analysis are promising. Making pairwise comparisons forces the decision maker to consider the importance of factors and to analyse the situation more precisely and in more depth. The applicability of the method in participatory planning will be studied in future. Public participation could be implemented by allowing all participants to perform their own SWOT analysis and pairwise comparisons and then aggregate separate results after weighting the participants according to individual importance. This might generate new alternatives and infuse more creativity in the planning process.

It is evident that a lot of managerial decision making is based on intuition and subjective judgements instead of the outcomes of formal planning. Expanding the presented formulation to cover a wider range of decision makers' and experts' input could benefit the planning process. Interaction, learning and consensus can all be achieved by, for example, including the Delphi technique in the planning process (e.g. Kangas *et al.* 1996). The hybrid method A'WOT increases and improves the information basis of strategic planning processes. It provides an effective framework for learning about strategic decision support in numerous situations. It can also be used as a tool in communication and education in decision-making processes where multiple decision-makers or judges are involved.

REFERENCES

AHOLA, J. 1995. Yrityksen strategiaprosessi; Lappeenrannan teknillinen korkeakoulu; Tieteellisiä julkaisuja no 44, Lappeenranta. Doctoral thesis.

ALHO, J. AND KANGAS, J. 1997. Analysing uncertainties in experts' opinions of forest plan performances. For. Sci. 43: 521-528.

ALHO, J., M., KANGAS, J. AND KOLEHMAINEN, O. 1996. Uncertainty in expert predictions of the ecological consequences of forest plans. Appl. Statist. 45: 1-14.

ANDREWS, K.R. 1971. The Concept of Corporate Strategy, Irwin, Homewood, IL.

ANDREWS, K.R. 1980. The Concept of Corporate Strategy, 2nd edition, Irwin, Homewood, IL.

ANSOFF, H. I. 1965. Corporate Strategy. McGraw-Hill, New York.

CAVES, R. E., MEHRA, S. K. 1986. Entry of Foreign Multinationals into U.S. Manufacturing Industries; Competition in Global Industries (edit. Michael E. Porter); Harvard Business School , Boston.

DESS, G. G. AND MILLER, A. 1993. Strategic Management. McGraw-Hill, New York.

FINCHEM, K. 1997. Embattled Paper Companies Seek Wise Capital Investment Strategies. Pulp and Paper International 1997: 51-57.

GEGINAT, H. 1998. Adapting to change will again be key for the future. Pulp and Paper International 1998: 3-3.

HERBERT, T. T. 1999. Multinational Strategic Planning: Matching Central Expectations to Local Realities. Long Range Planning: 81-87.

HIGHAM, R. 1996. Globalisation Strategies in Pulp, Paper and Board. Finacial Times, London.

HILL, T. AND WESTBROOK, R. 1997. SWOT analysis: It's time for a product recall. Long Range Planning 30: 46-52.

HOFER, C. W. AND SCHENDEL, D. 1978. Strategy Formulation: Analytical Concepts. West Publishing Co. St. Paul.

KANGAS, J., LOIKKANEN, T., PUKKALA, T. AND PYKÄLÄINEN, J. 1996. A participatory approach to tactical forest planning. Acta For. Fenn. 251. 24 pp.

KOTLER, P. 1988. Marketing Management: Analysis, Planning, Implementation, and Control. Sixth edition. Prentice-Hall International Edition.

KURTTILA, M., PESONEN, M., KANGAS, J. AND KAJANUS, M. 1999. Utilizing the analytical hierarchy process (AHP) in SWOT analysis - A hybrid method and its application to a forest-certification case. Forest Policy and Economics. In print.

KURTTILA, M., KANGAS, J., PESONEN, M., KAJANUS, M. AND HEINONEN, P. 1998. Using AHP and SWOT analyses in assessing priorities of strategy alternatives in forest planning. In: Saarikko, J. Resource Technology '98 Nordic. International Symposium on Advanced Technology in Environmental and Natural Resources, June 8-12, 1998, Rovaniemi, Finland. Abstracts. p. 16.

MCDONALD, M. H. B. 1993. The Marketing Planner. Butterworth-Heinemann, Oxford.

PAYNE, M. 1997. European Paper Company Strategies. Pulp an Paper International: 283-291.

PROCTOR, R. A. 1992. Selecting an appropriate strategy: A structured creative decision support model. Marketing Intelligence and Planning 10: 21 - 24.

RINNE, P. 1995. Concentrated effort changes Finnish industry structure. Pulp and Paper International 1995: 29-32.

SAATY, T. L. 1980. The Analytic Hierarchy Process, McGraw-Hill, New York.

VERY, P., LUBATKIN, M., CALORI, R. AND VEIGA, J. 1997. Relative Standing and the Performance of Recently Acquired European Firms. Strategic Management Journal 1997: 609-610.

WEIHRICH, H. 1982. The TOWS matrix: A tool for situational analysis. Long Range Planning 15(2): 54-66.

WHEELEN, T. L. AND HUNGER, J.D. 1995. Strategic Management and Business Policy, 5th Edition, Addison Wesley, Reading, MA.

Chapter 13

Prioritizing Salmon Habitat Restoration with the AHP, SMART, and Uncertain Data

Keith M. Reynolds
USDA Forest Service, Pacific Northwest Research Station, Corvallis, OR USA

Key words: Salmon habitat restoration, watershed, uncertainty, SMART, simple multi-attribute rating technique

Abstract: Ecological assessments provide essential background information about ecosystem states and processes and are thus a useful starting point for applying adaptive ecosystem management. As a logical follow-up to ecological assessment, managers may wish to identify, and set priorities for, ecosystem maintenance and restoration activities. The Simple Multi-Attribute Rating Technique (SMART) is a useful extension to the standard AHP model that allows characterisation of uncertainty in attribute values of alternatives, and thus is one way of incorporating risk analysis into the standard AHP model. Criterium DecisionPlus is used to demonstrate application of the AHP and SMART methods to the problem of evaluating priorities for salmon habitat restoration projects.

1. INTRODUCTION

Typical applications of the Analytic Hierarchy Process (AHP) include prioritising a set of alternatives to choose the best alternative from a set, or to use the priority scores as a basis for allocating limited resources among the alternatives (Golden *et al.* 1989). Earlier chapters describe the theory and principles of AHP. Several studies have described the general utility of the AHP as a decision support tool for forest planning (Kangas 1992, Pukkala and Kangas 1993, Kajanus *et al.* 1996, Kuusipalo *et al.* 1997). Some specific applications of the AHP in natural resource management include:

- evaluation of, and management for, biodiversity (Kangas and Kuusipalo 1993, Kuusipalo and Kangas 1994),

D.L. Schmoldt et al. (eds.),
The Analytic Hierarchy Process in Natural Resource and Environmental Decision Making, 199–217.

- habitat mapping (Steinmeyer *et al.* 1995),
- development of inventory and monitoring programs (Schmoldt *et al.* 1994),
- evaluation of factors contributing to risk of insect outbreaks (Reynolds and Holsten 1994), and
- setting priorities for maintenance and restoration projects (Reynolds 1997).

Ecological assessments provide essential background information about ecosystem states and processes and are thus a useful starting point for applying adaptive ecosystem management to management areas or regions. As a logical follow-up to ecological assessment, managers may wish to identify, and set priorities for, ecosystem maintenance and restoration activities. Decision models such as the Analytic Hierarchy Process and the Simple Multi-Attribute Rating Technique (SMART, Edwards 1977, Edwards and Newman 1982) provide a bridge between assessment and planning activities by helping managers to establish rational priorities for activities that may then subsequently inform the planning process.

Ecological assessments generally deal with a broad array of topics that include biophysical, social, and economic dimensions. Ideally, the same circumspection should carry over into processes used to identify, and set priorities for, maintenance and restoration activities derived from an assessment. AHP and SMART decision models are discussed together in this chapter because of their common ability to accommodate diverse types of decision criteria in a single model. SMART extensions to AHP allow characterisation of uncertainty in attribute values of alternatives, and thus allow the incorporation of risk analysis into the basic AHP model. Criterium DecisionPlus[2] (InfoHarvest, Redmond, WA) is used to demonstrate application of the AHP and SMART methods to the problem of evaluating priorities of maintenance and restoration projects.

2. EXTENDING THE AHP WITH SMART

SMART methods extend the AHP in two potentially useful ways (Kamenetzky 1982). First, SMART uses a utility function to map raw attribute values from an arbitrary numeric scale into the range [0, 1]. Typically, the numeric scale for attribute values corresponds to a range of verbal choices. For example, in the case of geographic scale, stand and

[2] The use of trade of firm names in this publication is for reader information and does not imply endorsement by the U.S. Department of Agriculture of any product or service.

regional projects might be assigned values of 0 and 100, respectively. The utility function may be derived from a statistical model if one exists, or it may be based on a qualitative heuristic model. The general form of the SMART utility function is

$$p_a^c = f^c(r_{min}^c, r_{max}^c, r_a^c) \tag{2.1}$$

in which p_a^c is the priority of alternative a with respect to lowest-level criterion c (also referred to as an attribute), f^c is the utility function for attribute c that maps the alternative value r_a^c onto the range [0, 1], and r_{min}^c and r_{max}^c are parameters that define the shape of f^c.

Integration of the SMART utility function into an AHP model is straightforward. For example, for a hierarchy with m levels of criteria, the decision score for alternative a can be represented as

$$d_a = \sum_{c1=1}^{n^g} W_{c1}^g \sum_{c2=1}^{n^{c1}} W_{c2}^{c1} \ldots \sum_{cm=1}^{n^{cm-1}} W_{cm}^{cm-1} p_a^{cm} \tag{2.2}$$

in which d_a is the decision score of alternative a, the W_*^* are the usual criterion weights from the AHP, and the p_a^{cm} are the priorities of alternative a with respect to all lowest-level criteria (2.1). Formula (2.2) expresses that the decision score is calculated as the alternative priority with respect to each lowest-level criterion times the accumulated weight of that criterion, summed over all lowest-level criteria. Comparing formula (2.2) to the standard AHP model, formula (2.2) simply substitutes formula (2.1) for a final weight term that either is derived from pairwise comparisons between alternatives with respect to each lowest-level criterion, or is given directly (that is, by direct ranking of alternatives on a [0, 1] scale).

The second useful feature of the SMART method lets the analyst specify a statistical distribution around expected attribute values for criteria associated with specific alternatives to express uncertainty about the accuracy of attribute values entering an analysis. For example, actual costs of prospective projects are rarely, if ever, perfectly known. In SMART, the analyst can specify both an expected value for cost and the distribution of cost about its expected value. Uncertainty about cost is then translated into uncertainty about the utility function. Ultimately, all sources of uncertainty relevant to an alternative under consideration are expressed in cumulative uncertainty about the computed priority for the alternative. The general probability density $\rho_Z(Z)$ of a composite variable $Z=\Sigma x_i$ is given by the generally intractable integral

$$\rho_Z(Z) = \int dx_1 x_2 ... x_n P(x_1, x_2, ..., x_n) \delta(Z - \sum_{i=1}^{n} x_i) \tag{2.3}$$

in which $P(x_1, x_2, ..., x_n)$ is the joint probability density of the x_i and $\delta(w)$ is a delta function that is only nonzero when $w = 0$. In the Criterium DecisionPlus implementation of AHP and SMART, the x_i (that is, the attribute scores of an alternative with respect to all lowest-level criteria) are assumed to not be correlated, in which case the joint probability distribution simplifies to

$$P(x_1, x_2, ..., x_n) = \rho_1(x_1)\rho_2(x_2)...\rho_n(x_n) \tag{2.4}$$

and the integral (2.3) can then be expressed as a series of iterated convolutions

$$p_Z(Z) = \int dx_1 \rho_1(x_1) \int dx_2 \rho_2(x_2) ... \int dx_{n-1} \rho_{n-1}(x_{n-1}) \rho_n(Z - \sum_{i=1}^{n-1} x_i)) \tag{2.5}$$

based on properties of the delta function. If we rewrite formula (2.2) as

$$d_a = \sum_{l=1}^{n} W^l_{accum} p_l(s^a_l) \tag{2.6}$$

in which W^l_{accum} is now the accumulated weight on lowest-level criterion l, s^a_l is the score of alternative a with respect to l, and let $(x^a_l) = W^l_{accum} p_l(s^a_l)$, then $\rho_{d_a}(d_a) = \sum_l x^u_l$, and the probability density over x^a_l is calculable as

$$\rho_l(x^a_l) = \frac{\partial s^a_l}{\partial x^a_l}(s^a_l)\rho_{s^a_l}(s^a_l) \tag{2.7}$$

Thus, SMART provides a basis for evaluating the risk of incorrectly choosing the alternative with highest priority, based on known or estimated measurement errors in attribute values, but cumulative error estimates are conservative in the sense that they are upper bounds on the true errors.

3. ANALYSIS CONTEXT

Reynolds and Reeves (in press) evaluated 6th-code hydrologic units (hereafter, watersheds) in the Nestucca Basin of the northern Oregon Coast

Range for salmon habitat suitability. Sixth-code watersheds delineate drainage basins, which, in the Western US, typically range in area from about 10,000 to 20,000 hectares. Evaluations of biophysical attributes of the 6th-code watersheds in the study area resulted in conclusions of slight to severe reductions in habitat suitability for most units (Figure 1). Frequencies with which environmental conditions substantially contributed to a conclusion of reduced suitability were separately summarised by Reynolds and Reeves (in press) for units with high- and low-gradient stream reaches because different evaluation criteria applied in the two cases.

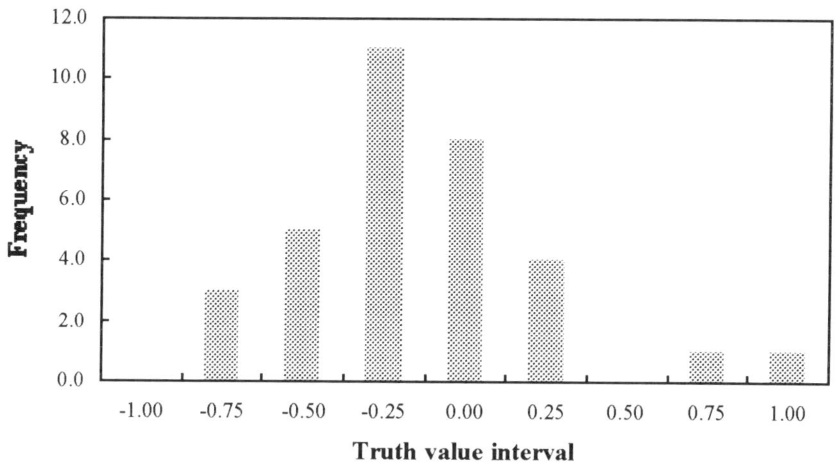

Figure 1. Frequency distribution of the salmon habitat suitability index in the Nestucca Basin analysis.

Given the results of this particular ecological assessment, two questions come to mind. Which watersheds are most in need of restoration? And, which impaired watersheds should be the highest priority for restoration activities? Although these two questions are similar, they are not the same. To answer the first, it is sufficient to consider the results of the analysis (Figure 1). Clearly, there are several watersheds in the Nestucca Basin that evaluated as having rather low salmon habitat suitability. However, in considering which watersheds should receive highest priority for restoration, decision criteria related to project feasibility and efficacy also are relevant to managers.

Many factors pertinent to evaluating habitat suitability also are relevant to considerations of feasibility and efficacy. For example, road density, number of road crossings over landslide-prone areas, and amount of farm activity within a watershed are all factors that influence salmon habitat suitability in high-gradient watersheds. Sediment delivery to streams and stream siltation both tend to increase with road density, with consequent reductions in salmon habitat suitability. Likewise, frequency of road crossings over landslide-prone areas is associated with increased sedimentation and siltation, while farm activity has a bearing on stream bank condition and water quality. All three factors also are relevant to the feasibility of restoration. Roads are expensive to decommission, and eliminating or significantly modifying deleterious farm practices may not be politically feasible for a variety of reasons.

Similarly, reduced upland forest cover in mature age classes is associated with reduced habitat suitability, and reduced extant upland forest cover may reduce the efficacy of road decommissioning actions because effects of reduced cover include increased sediment delivery and higher stream temperature. The general conclusion to be drawn from these last few points is that many factors associated with evaluation of habitat suitability also are potentially pertinent to decisions about remediation because these same factors are relevant to considerations of restoration feasibility and efficacy. Thus, ecological assessments and subsequent decisions about restoration projects may be more or less closely coupled, depending on the extent to which the two phases of analysis share attributes.

4. THE DECISION HIERARCHIES

Two separate AHP decision hierarchies were designed to prioritise watersheds for salmon habitat restoration in the Nestucca Basin because details of the ecological assessment varied, depending on whether a watershed had a high or low stream reach gradient. The decision hierarchy for watersheds with high gradients (Figure 2) is simpler than that for low gradients (Figure 3) because in-channel stream attributes were not relevant to the ecological assessment of high-gradient watersheds (Reynolds and Reeves, in press). However, both decision hierarchies share the same set of primary criteria: Habitat suitability, Critical habitat, Feasibility, and Efficacy (Table 1). Pairwise comparisons were performed among primary criteria (Table 2) to derive weights for the relative importance of each criterion (Golden *et al.* 1989, Saaty 1992), and Reeves and Hohler (USDA Forest Service, Pacific Northwest Research Station, personal communication) validated these judgements. The consistency ratio for comparisons of primary criteria was 0.002, indicating a very high level of consistency in pairwise judgements (Golden *et al.* 1989, Saaty 1992).

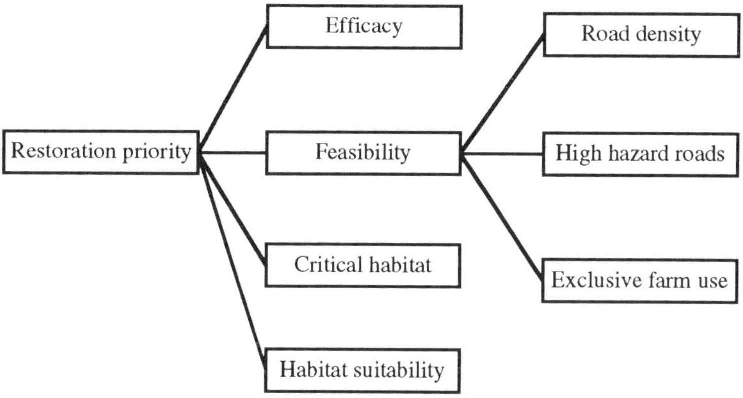

Figure 2. Decision hierarchy for prioritising salmon habitat restoration in 6th-code watersheds with high stream reach gradients.

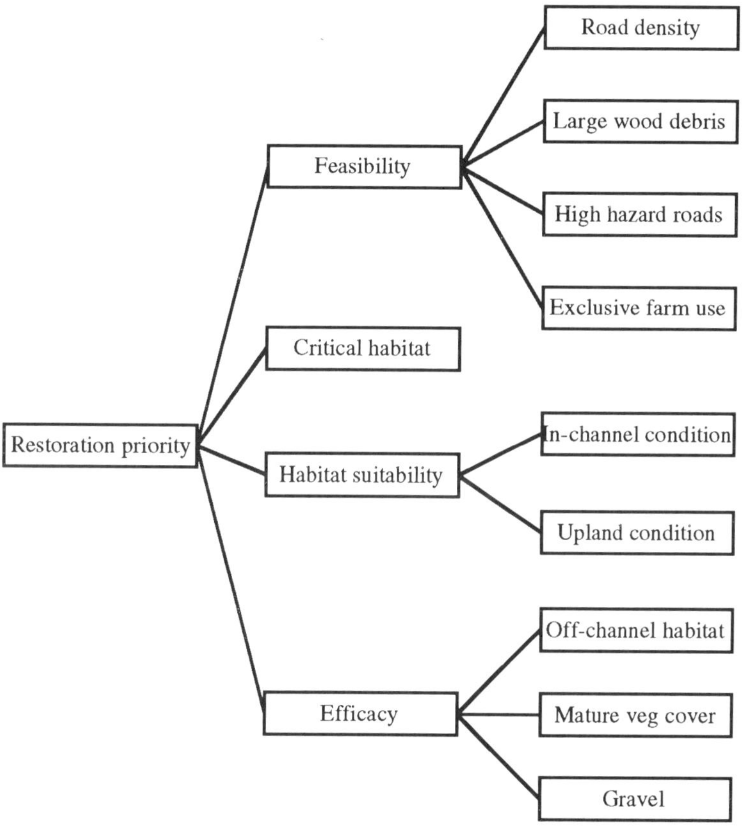

Figure 3. Decision hierarchy for prioritising salmon habitat restoration in 6th-code watersheds with low stream reach gradients.

Table 1. Descriptions of primary AHP criteria for prioritising watershed restoration.

Criterion	Description	Attributes or subcriteria[a]
High-gradient watersheds[b]		
	Condition of habitat warrants restoration.	Habitat suitability index.[c]
	Watershed provides critical salmon habitat within the basin.	Exogenous variable provided by author.
	Watershed attributes are conducive to restoration.	Percent of upland forest cover in mature age class (80-150 years old).
	Restoration is cost-effective and politically feasible.	Road density (miles per square mile).
		Road crossings over landslide-prone areas per mile perennial stream.
		Percent area designated for exclusive farm use.
Low-gradient watersheds[d]		
	Condition of habitat warrants restoration.	Upland condition index.[c]
		In-channel condition index.[c]
	Watershed provides critical salmon habitat within the basin.	Exogenous variable provided by author.
	Watershed attributes are conducive to restoration.	Percent upland forest cover in mature age class (80-150 years old).
		Percent stream surface area with suitable off-channel habitat for salmon.
		Percent gravel cover in riffle substrate.
	Restoration is cost-effective and politically feasible.	Road density (miles per square mile).
		Number of road crossings over landslide-prone areas per mile perennial stream.
		Percent area designated for exclusive farm use.
		Number of pieces of large woody debris per mile of stream.

[a] Lowest level criteria in a decision hierarchy are also attributes.
[b] High-gradient watersheds have mean stream reach gradient greater than 4 percent.
[c] From Reynolds and Reeves (in press).
[d] Low-gradient watersheds have mean stream reach gradient less than or equal to 4 percent.

4.1 SubCriteria for High-Gradient Watersheds

Only the Feasibility criterion had subcriteria in the hierarchy for high-gradient watersheds (Figure 2, Table 1). Weights for the relative importance of subcriteria of Feasibility were derived from pairwise comparisons (Table 3), for which the consistency ratio was 0.021, again indicating a high level of consistency in pairwise judgements.

Table 2. Pairwise comparisons of the relative importance of criteria contributing to restoration priority of high- and low-gradient 6th-code watersheds in the Nestucca Basin.[a]

	Criteria			
Criteria	Habitat suitability	Critical habitat	Efficacy	Feasibility
Habitat suitability	1	1	4	4
Critical habitat	1	1	4	5
Efficacy	-	1/4	1	1
Feasibility	-	1/5	1	1

[a] Numeric ratings correspond to the standard 9-point comparison scale of the AHP.

4.2 SubCriteria for Low-Gradient Watersheds

The Habitat suitability, Feasibility and Efficacy criteria each had sub-criteria in this hierarchy (Figure 3, Table 1). Weights for relative importance of subcriteria of Feasibility and Efficacy were derived from pairwise comparisons (Table 4 and 5, respectively), for which the consistency ratios were 0.018 and 0.033, respectively, again indicating high levels of consistency in pairwise judgements.

5. EVALUATION OF ALTERNATIVE ATTRIBUTES WITH SMART

Attributes of all alternatives in both decision hierarchies were evaluated with SMART. The three index values (Habitat suitability in Figure 2 and Upland condition and In-channel condition in Figure 3) were normalised to the 0-1 SMART utility scale by simple linear transformations. For most other attributes, the response scale for the attribute was defined to correspond to a corresponding fuzzy membership function definition from Reynolds and Reeves (in press). Similar to utility functions, fuzzy membership functions evaluate propositions about data by mapping data (or attribute) values onto a scale that expresses the value's degree of membership in a set. The form of a fuzzy membership function is more or less arbitrary, subject only to the requirement that the function maps values to some standard scale that expresses fuzzy set membership. Attribute values were transformed to the [0-1] utility scale with exponential functions parameterised so that an attribute value returning a fuzzy membership value of 0 (on a [–1, 1] fuzzy membership scale) mapped to a utility value of 0.5.

Table 3. Pairwise comparisons of the relative importance of subcriteria contributing to restoration feasibility of high-gradient 6th-code watersheds in the Nestucca Basin.[a]

	Criteria		
Criteria	Exclusive farm use	High hazard roads	Road density
Exclusive farm use	1	2	5
High hazard roads	1/2	1	4
Road density	1/5	1/4	1

[a] See footnote for Table 2.

Table 4. Pairwise comparisons of the relative importance of subcriteria contributing to restoration feasibility of low-gradient 6th-code watersheds in the Nestucca Basin.[a]

	Criteria			
Criteria	Exclusive farm use	High hazard roads	Large woody debris	Road density
Exclusive farm use	1	2	1	5
High hazard roads	1/2	1	1	4
Large woody debris	1	1	1	4
Road density	1/5	1/4	1/4	1

[a] See footnote for Table 2.

Table 5. Pairwise comparisons of the relative importance of subcriteria contributing to restoration efficacy of low-gradient 6th-code watersheds in the Nestucca Basin.[a]

	Criteria		
Criteria	Gravel substrate	Off-channel habitat	Mature veg cover
Gravel substrate	1	3	5
Off-channel habitat	1/3	1	3
Mature veg cover	1/5	1/3	1

[a] See footnote for Table 2.

As an example of the correspondence between SMART attribute scale definitions and fuzzy membership functions and the mapping of attribute values into utility functions, consider the fuzzy membership function definition for evaluation of Mature vegetation cover used in Reynolds and Reeves (in press). The function for evaluating fuzzy membership of an observation on Mature vegetation cover defines values of 30 percent or less as not suitable for salmon habitat, defines cover values of 45 percent or more as completely suitable, and defines values between 30 and 45 percent as having partial suitability (Figure 4). To maintain a simple relation between the fuzzy membership and SMART utility functions, the SMART attribute response range for Mature vegetation cover was defined on the closed interval [30, 45], and observed values outside this range were set to the appropriate minimum or maximum condition (Figure 5). Notice that a value of 37.5 percent Mature veg cover has a fuzzy membership value of 0 (Figure 4). Because fuzzy membership in our application is defined on the closed interval [-1, 1], a fuzzy membership value of 0 corresponds to 50 percent

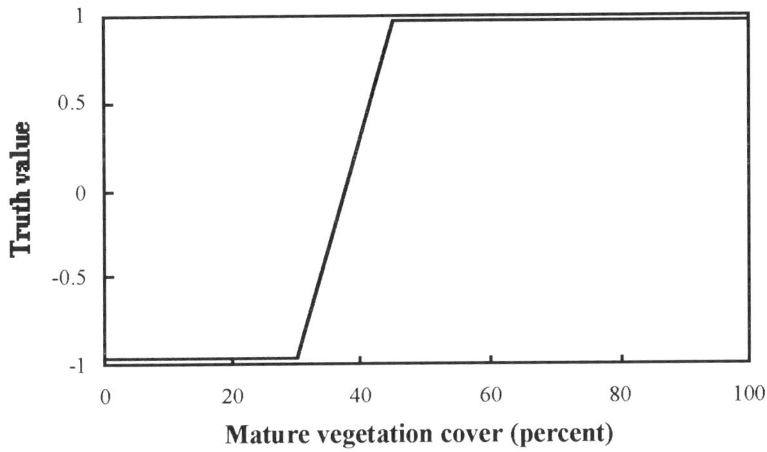

Figure 4. Fuzzy membership function for evaluating the proposition that observed Mature veg cover is suitable for salmon habitat.

membership on the more usual [0, 1] scale. The exponential utility function for evaluation of Mature veg cover was parameterised in SMART so that an observed value of 37.5 percent evaluated to a utility score of 0.5 (Figure 5).

$$u_{Matureveg} = e^{-0.0185 v_{mature}} \tag{5.1}$$

The objective in formulating utility functions in this manner was to maintain a set of consistent relations between evaluation of attribute values used to assess habitat condition and the SMART utility functions used to evaluate restoration priority.

Attribute data for analyses in general are rarely measured with complete accuracy. In the case of AHP analyses in particular, attribute measurement errors can result in miscalculation of priorities, increasing the risk of selecting sub-optimal alternatives. The ability to specify error distributions for observations on any attributes of any alternative is a useful capability of the SMART methodology as implemented in Criterium DecisionPlus (Figure 6). In the analysis (next section), error estimates are propagated upward through the AHP decision hierarchy to produce an integrated assessment of error for each alternative's priority. To illustrate how SMART integrates data errors in the AHP model in this analysis, reasonable estimates of measurement error were provided for the attributes, Mature vegetation cover, Large woody debris, Off-channel habitat, and Gravel, by assuming a 10 percent standard error around the observed value, assuming normal error distributions. Index values from the ecological assessment (Habitat

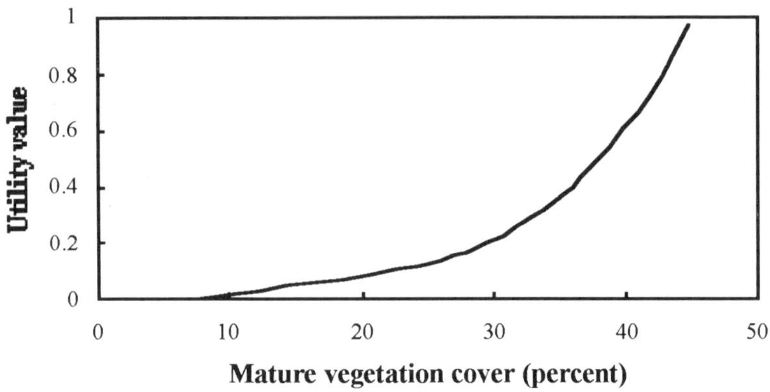

Figure 5. SMART utility function relating observed values of Mature veg cover to the Efficacy criterion for habitat restoration.

suitability in Figure 2 and In-channel condition and Upland condition Figure 3) were assumed to be calculated with a uniformly distributed error of ±0.1.

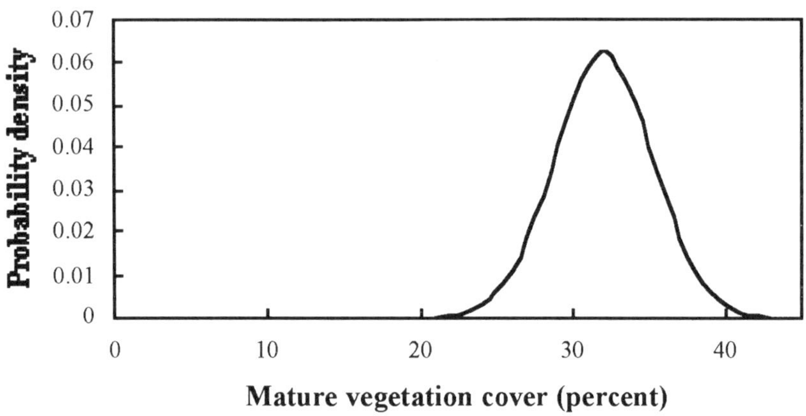

Figure 6. Specifying a measurement error for Mature veg(etation) cover for the Alder Creek alternative.

Uncertainties may enter an analysis in other ways. For example, in the present context, there are data on road densities and frequencies of roads crossing landslide-prone areas. One can assume that these data contain a high degree of accuracy, so measurement errors are not a primary concern. On the other hand, there were no road engineering studies available, so the length of road that would actually need to be decommissioned to improve salmon habitat was not known. To account for this uncertainty, I specified a normal error distribution for each road attribute value with mean = $x/2$ and standard deviation set to approximately $x/6$ so that the error distribution spanned the closed interval $[0, x]$, where x is the observed road length.

6. RESTORATION PRIORITIES

6.1 High-Gradient Watersheds

The North Beaver watershed achieved the highest priority score in the analysis of high-gradient watersheds, and always rated highest even when all sources of attribute data uncertainty were taken into account (Figure 7). Similarly, the watershed with the next highest rating (Little Beaver) always rated as second in priority for restoration. Note, however, that the next group of four watersheds (Walker Creek through Slickrock Figure 7) is not only tightly clustered in terms of priority ratings, but there is considerable overlap in error bars associated with priority scores. Within this group of four watersheds, there is a nontrivial risk of incorrectly giving higher priority to one over the others, and the risk is particularly significant for the two pairs, Walker Creek versus McGuire Reservoir and West versus Slickrock.

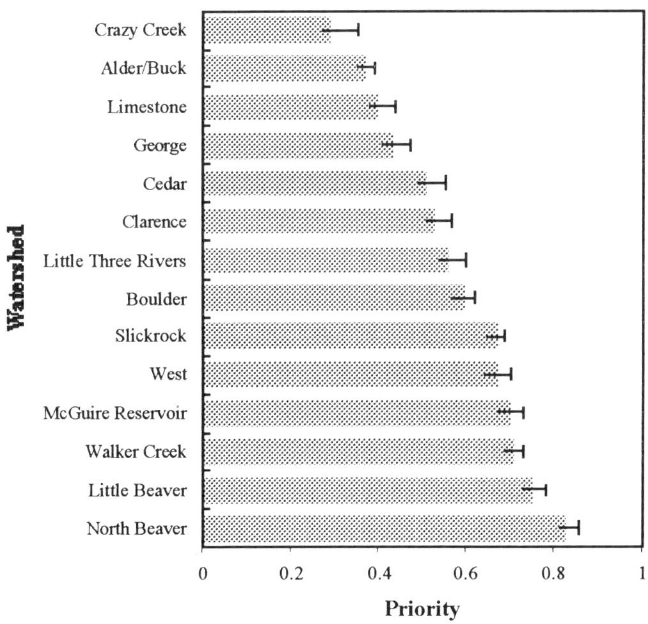

Figure 7. Restoration priorities for high-gradient watersheds.

The basis for the high priority of North Beaver can be seen in the contributions of criteria to priorities of the five most highly rated watersheds (Figure 8). North Beaver rated very high with respect to both Critical habitat and Habitat suitability compared to any of the next four alternatives. While the next four watersheds have similar total overall priorities, there is considerable variation in the factors contributing to the priority rating. In all five cases (Figure 8), Efficacy makes only a modest contribution to the overall rating. This is consistent with sensitivity analyses that indicated that alterations in ratings were most sensitive to the Efficacy criterion (Figure 9). However, a proportional change of 16.9% in weight of the Efficacy criterion would be needed to produce a re-ordering of priorities. Such a magnitude of change is substantial, and, because Efficacy is the most sensitive of all criteria, it can be concluded that the analysis is reasonably robust with respect to weights derived from pairwise comparisons among criteria, given observed attribute values of the alternatives.

6.2 Low-Gradient Watersheds

The Wolfe watershed ranked highest in restoration priority in the analysis of low-gradient watersheds, but given the sources of error in attributes of alternatives, it only ranks as best 92% of the time (Figure 10). A value of 92% is quite high, however, so Wolfe could be selected as the highest priority with low risk of making an inappropriate choice. The next two highest rated alternatives (Tiger and Farmer) achieve almost the same

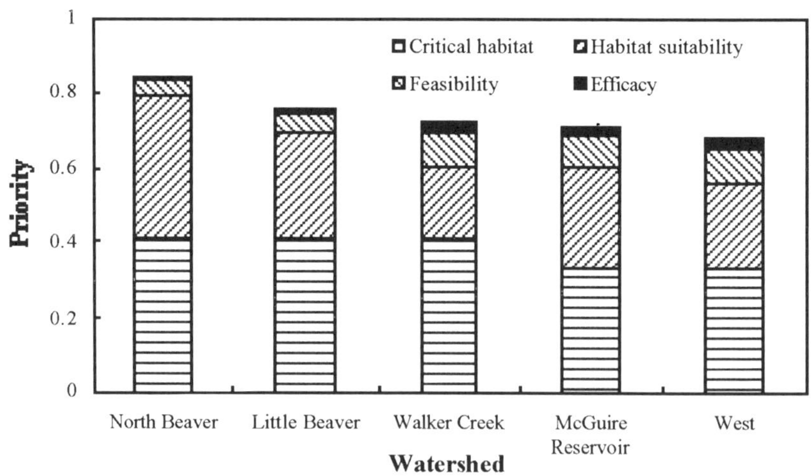

Figure 8. Contributions of primary criteria to priority rating of the five most critical high-gradient watersheds.

priority ratings, have similar error distributions, but also are fairly clearly separable from the next highest rated alternative when considering both computed priority and estimated error. Overall priorities for the three highest ranked alternatives were quite close to one another, but as in the case of the high-gradient analysis, the highest-ranking alternative (Wolfe) edges out the next two largely based on high contributions from both Habitat suitability and Critical habitat (Figure 11). Comparing the second and third alternatives, Tiger outranks Farmer based on higher Efficacy for the Tiger alternative. Sensitivity analyses again indicated that priority rank was most sensitive to change in the weight on Efficacy. In this case, a change in Efficacy weight of 4.4% would result in a re-ordering of priorities (Figure 12), and indicates that weighting of Efficacy compared to other primary criteria might warrant closer attention. Priority rankings were also moderately sensitive to changes in weights on In-channel condition and Upland condition (8.7% change in both cases).

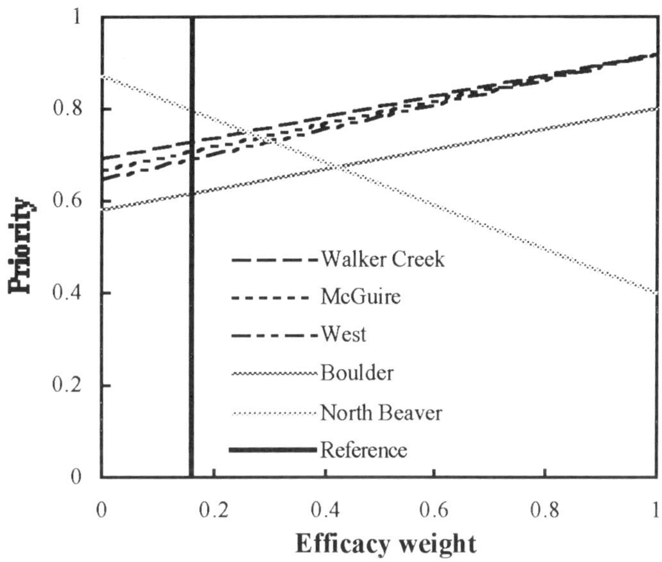

Figure 9. Sensitivity of priority rating to the Efficacy criterion in the analysis for high-gradient watersheds.

7. DISCUSSION

The example analysis presented in this chapter only involved a limited number of watersheds, but the AHP model for salmon habitat restoration was developed for the more general situation in which 50 to 100 alternative watersheds might commonly need to be considered in a given analysis. In standard AHP analysis, ratings on attribute values are derived by pairwise comparisons between alternatives with respect to lowest-level criteria, or ratings are directly assigned (Saaty 1994). Direct assignment of priority ratings to attribute scores commonly is used when the number of alternatives in a given AHP model varies over time, or when the number of alternatives to consider makes pairwise comparisons between alternatives impractical. Use of the SMART extension to AHP is worth consideration for quantitative attributes when the number of alternatives being considered is large enough that direct ranking would be preferable to pairwise comparisons between alternatives because SMART also accommodates error propagation from attribute values to priorities for alternatives.

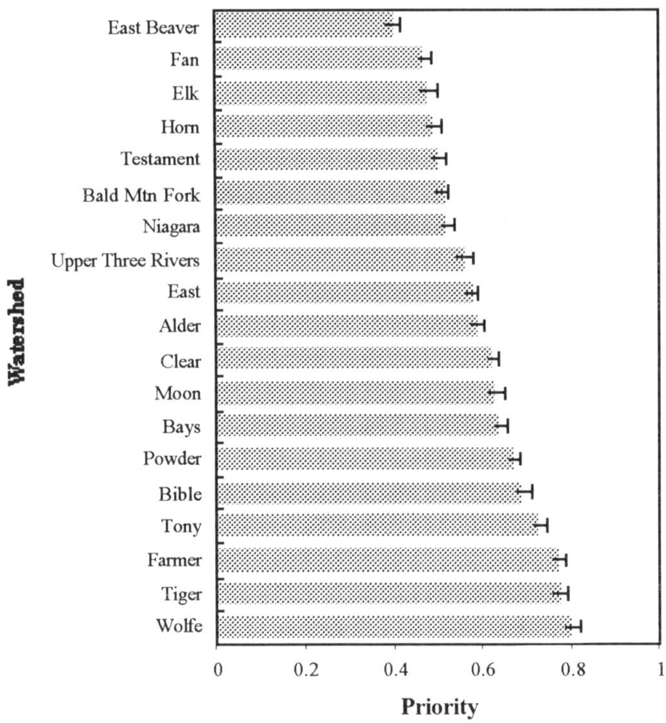

Figure 10. Restoration priorities for low-gradient watersheds.

The AHP model for salmon habitat restoration treats criterion weights as deterministic. This was acceptable in our situation because there were only two experts involved in developing the model and they had no difficulty arriving at criterion weights by consensus. More generally, probability density distributions for priority of alternatives are affected not only by error distributions for attribute values, but also by error distributions for criterion weights developed by judgements (Saaty 1994). Some AHP applications, such as Criterium DecisionPlus, handle the former error problem while others, such as Expert Choice (Pittsburgh, PA), handle the latter. I am not aware of any currently available commercial applications, however, that simultaneously treat both sources of error. This applies to standard AHP models where weights for attributes are developed by pairwise comparisons because errors in attribute weights derived from judgements and errors in measuring attribute values constitute two distinct sources of error.

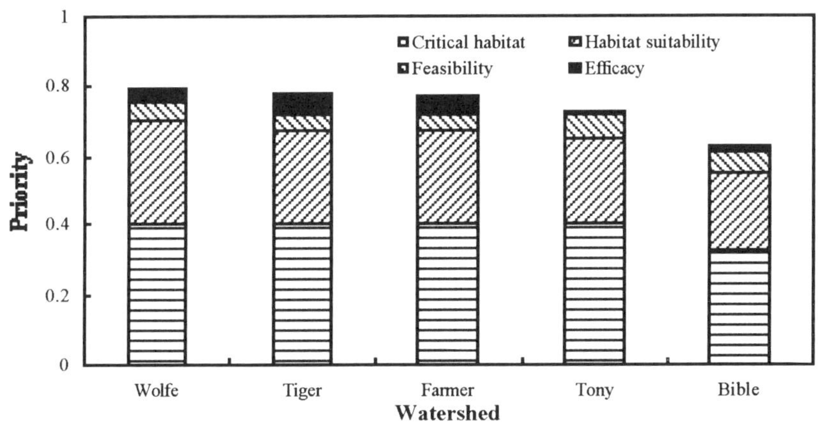

Figure 11. Contributions of primary criteria to priority rating of the five most critical low-gradient watersheds.

A recurrent theme in this chapter has been the possibility of achieving some degree of integration between the different phases of analysis related to resource management topics such as habitat suitability. Any number of other topics could have served equally well. For example, it is easy to conceive of approaching conservation of biodiversity in a way very similar to what has been presented here. To the extent that one analysis flows naturally from the other, the analyst is building a logical trail in which each prior step supports the succeeding one, so that, in the end, there is a clear logical path from the question, "What is the state of this system?", to the answer, "Here's how to respond", and the justification, "This is why we should respond in this way." A few motivating questions provide the basis for the approach that I have illustrated:

- What information contributed to the conclusions about the state of the resource?
- Is any of this information relevant to decisions about how to respond?

- In particular, is any of the information useful as context that could influence considerations of efficacy or feasibility of implementing a response? and
- Is there any other information not needed to evaluate the state of the resource, but relevant to setting priorities, etc.?

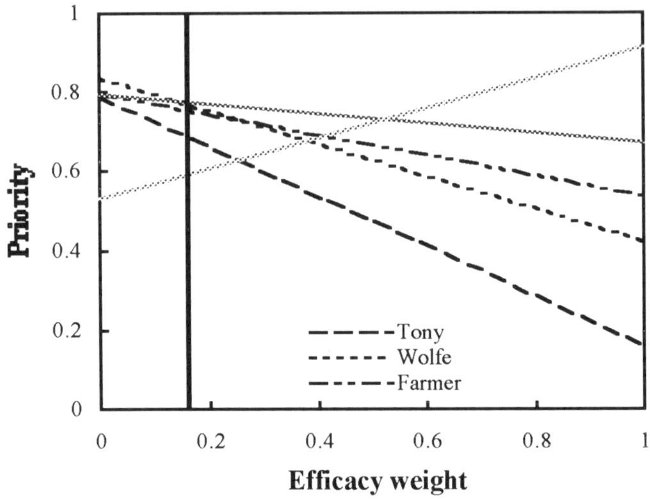

Figure 12. Sensitivity of priority rating to the Efficacy criterion in the analysis for low-gradient watersheds.

Finally, the AHP is a powerful analytical tool for decision making tasks in natural resource management that involve setting priorities, selecting alternatives, or allocating resources. Key to the continued growth in its popularity, the AHP provides a rational formalism for problem representation that is both easy to apply and easy to communicate to interested parties. SMART usefully extends the functionality of the basic AHP methodology by providing a simple interface for quickly and easily normalising raw attribute values entering an analysis. The implementation of SMART in Criterium DecisionPlus provides a good example of how the basic AHP approach can be extended further still with the incorporation of error estimates for risk analysis in an AHP context.

REFERENCES

EDWARDS, W. 1977. How to use multi-attribute utility measurement for social decision making. IEEE Transactions on Systems, Man, and Cybernetics 7:326-340.

EDWARDS, W., AND NEWMAN, J. R. 1982. Multi-attribute Evaluation. Sage, Beverly Hills, CA.

GOLDEN, B. L., HARKER, P. T. AND WASIL, E. A. 1989. The Analytical Hierarchy Process: Applications and Studies. Springer-Verlag, New York.

KAJANUS, M., KURTTILA, M., PESONEN, M., HYTTINEN, P. AND NILSON, A. 1996. Applying SWOT and AHP analysis when changing to ecolabelled forestry. Pages 217-226 in EFI Proceedings: Integrating Environmental Values into Forest Planning - Baltic and Nordic Perspectives. Proceedings of the Nordic-Baltic Research Course, Rapina, Estonia, 2 June 1927-2 July 1996. European Forest Institute, Joensuu, Finland.

KAMENETZKY, R. 1982. The relationship between the analytical hierarchy process and the additive value function. Decision Sciences 13:702-716.

KANGAS, J. 1992. Multiple-use planning of forest resources by using the analytic hierarchy process. Scandinavian Journal of Forest Research 7:259-268.

KANGAS, J. AND KUUSIPALO, J. 1993. Integrating biodiversity into forest management planning and decision-making. Forest Ecology and Management 61:1-15.

KUUSIPALO, J. AND KANGAS, J. 1994. Managing biodiversity in a forestry environment. Conservation Biology 8:450-460.

KUUSIPALO, J., KANGAS, J. AND VESA, L. 1997. Sustainable forest management in tropical rain forests: a planning approach and case study from Indonesian Borneo. Journal of Sustainable Forestry 5:93-118.

PUKKALA, T. AND KANGAS, J. 1993. A heuristic optimization message for forest planning and decision making. Scandinavian Journal of Forest Research 8:560-570.

REYNOLDS, K. 1997. Setting priorities for maintenance and restoration projects with the Analytic Hierarchy Process and SMART. Pages 2089-2095 in Resource Technology 97 Conference Proceedings, Seattle, WA, 7 April 1997-10 April 1997. American Society for Photogrammetry and Remote Sensing, American Congress on Surveying and Mapping, Bethesda, MD.

REYNOLDS, K. M. AND HOLSTEN, E. H. 1994. Relative importance of risk factors for spruce beetle outbreaks. Canadian Journal of Forest Research 24:2089-2095.

REYNOLDS, K. M. AND REEVES, G. H. (in press). Analysis and communication of monitoring data with knowledge-based systems. In Views from the Ridge Conference Proceedings. Gen. Tech. Rep. PNW. USDA Forest Service, Pacific Northwest Research Station, Portland, OR.

SAATY, T. L. 1992. Multicriteria Decision Making: The Analytical Hierarchy Process. RWS Publications, Pittsburgh.

SAATY, T. L. 1994. Fundamentals of Decision Making and Priority Theory with the Analytic Hierarchy Process. RWS Publications, Pittsburgh.

SCHMOLDT, D. L., PETERSON, D. L. AND SILSBEE, D. G. 1994. Developing inventory and monitoring programs based on multiple objectives. Environmental Management 18:707-727.

STEINMEYER, A., GADOW, K. V. AND VON, G. K. 1995. Saaty's AHP, as exemplified in forest habitat mapping. Centralblatt fur das Gesamte Forstwesen 112:53-65.

Chapter 14

A Fuzzy Analytic Hierarchy Process for Assessing Biodiversity Conservation

G. A. Mendoza
Department of Natural Resources and Environmental Sciences, University of Illinois, Urbana, Illinois USA

R. Prabhu
Centre for International Forestry Research (CIFOR) Regional Office, Institute of Environmental Studies, University of Zimbabwe, Harare, Zimbabwe

Key words: Multi-criteria decision making, fuzzy sets, sustainable forest management, biodiversity, criteria and indicators of sustainability

Abstract: Biodiversity conservation, as one criterion for evaluating forest sustainability, is assessed using a fuzzy analytic hierarchy process. Biodiversity conservation is presented in a hierarchical framework organised around three concepts: criteria, indicators, and verifiers. Following this hierarchical framework, a biodiversity conservation index model was developed as a two-step process. First, analysis is done at the indicator level to estimate the cumulative impacts of the verifiers, which are modelled as fuzzy variables. At the second level, the cumulative impacts of the indicators are measured. In both levels, the analytic hierarchy process is used to estimate the relative importance of each element in the hierarchy. Biodiversity experts then provided opinions, through pairwise comparisons, for elements at each level in the hierarchy, producing estimates of their relative importance. Based on these importance values, a composite biodiversity conservation index is calculated by combining index values from both levels of the hierarchy. To demonstrate this approach, a case study involving a forest located in Kalimantan, Indonesia was used.

1. INTRODUCTION

Over the last decade, concern about the sustainable management of remaining global forests has received word-wide attention. Tropical forests,

D.L. Schmoldt et al. (eds.),
The Analytic Hierarchy Process in Natural Resource and Environmental Decision Making, 219–233.

in particular, have received world-wide concern because of their urgent need for sustainable management. In response to this need, several national and international initiatives have been adopted and implemented. One such initiative is the development of criteria and indicators for assessing sustainable forest management. Mendoza and Prabhu (2000) examines some important issues related to measurement and assessment of forest sustainability based on four conceptual tools, namely: Principles, Criteria, Indicators, and Verifiers. These conceptual tools, generically referred to as Criteria and Indicators (C&I) in the literature, are proposed as instruments to measure forest sustainability.

Along with sustainable forest management, other concerns like biodiversity conservation have also gained currency in the forest management literature. Biodiversity, in general, is a topic that has received worldwide attention among forest and natural resource professionals and environment-oriented organisations because of the widely perceived problem of species extinction in some areas.

The general purpose of this paper is to examine biodiversity conservation as one criterion for evaluating forest sustainability. For this purpose, the paper describes the use of formal procedures to carry out the analysis. Specifically, the paper proposes a fuzzy methodology based on the Analytic Hierarchy Process (Saaty 1995) to structure the analysis of biodiversity conservation.

2. BIODIVERSITY AND BIODIVERSITY CONSERVATION

The popularity of the term biodiversity (a contraction of biological diversity) has increased dramatically over the last decade. Despite some of the confusion about its exact meaning (e.g., genetic diversity, species diversity, ecosystem diversity), several initiatives have been undertaken to mitigate the rapid decline of species, especially in areas where their habitat is threatened and the species themselves are endangered or even close to extinction.

Biodiversity conservation, sometimes referred to as conservation biology, is one of the strategies adopted to alleviate perceived biodiversity problems. The primary goal of conservation biology is the preservation of biological diversity not only through genetic and breeding approaches, but more importantly through the protection of natural areas and habitats. Hence, rather than protecting individual species, establishing gene banks, or pursuing breeding programs for individual species, a more holistic approach that works in both ex situ and in situ contexts must be adopted. This essentially is the approach of biodiversity conservation.

3. WHY USE THE AHP IN BIODIVERSITY CONSERVATION ANALYSIS?

Assessing biodiversity conservation is inherently a complex undertaking. While biodiversity itself may have a precise definition and meaning, biodiversity conservation is more difficult to define universally. This difficulty arises due to its broad scope not only in terms of spatial scales, but also in terms of the different physical and biological factors it encompasses, including human-induced effects. Because of its holistic nature and the broad attributes it embraces, biodiversity conservation is deemed a suitable criterion for forest sustainability assessments.

The concept of biodiversity conservation encapsulates several factors each with its own unique attributes. Many of these factors may not be easily identifiable or amenable to direct measurement or quantification. Consequently, assessing biodiversity conservation can be problematic, especially if traditional evaluative tools are used. Invariably, the multi-faceted nature of biodiversity conservation, its wide spatial scale and the multiple issues it encompasses defy attempts to analyse it using precise and more exacting methodologies. On the other hand, strictly ad hoc procedures increase the possibility of generating questionable or contestable assessments. Such unfavourable occurrence may be exacerbated by informal assessment procedures because they offer little or no "track record" that helps explain the rationale or logic employed. This and the lack of transparency of ad hoc assessment processes can hinder acceptance of the biodiversity conservation analysis.

AHP offers a convenient framework for biodiversity conservation analysis. The four-step process of AHP as described in Mendoza and Prabhu (2000) provides a structured approach that enables systematic evaluation of the factors and issues encompassed within biodiversity conservation. The first step consists of de-constructing or decomposing biodiversity conservation into a multi-level hierarchy. The hierarchy consists of a variety of elements that operate at different levels. Within each level, different factors, or indicators in the context of C&I assessments, can be identified.

The next step is assessment, via pairwise comparison, of the comparative importance of the different indicators. As described by Mendoza and Prabhu (2000), this step provides the basic information that will be used to estimate the relative importance of individual indicators or factors. In this study, pairwise comparisons were based on the ratio scale proposed by Saaty

(1995). In general, the pairwise comparisons are expressed on a scale between 1 (denoting equal importance) to 9 (denoting absolute importance). Intermediate scales between 1 and 9 denote varying degrees of importance from weak to extreme.

Synthesis of pairwise comparisons constitutes the third step. The result of this step is the calculated relative weights of individual indicators reflecting their relative importance. Finally, the fourth step consists of prioritising the list of indicators based on their estimated relative importance values.

In addition to the four-step process described above, AHP also has some desirable characteristics that make it an appropriate tool for assessing biodiversity conservation. Firstly, AHP can accommodate multiple experts in the assessment process. Secondly, it can incorporate mixed data that may include both qualitative and quantitative judgements. Thirdly, it is capable of analysing multiple factors, both individually and collectively. These features help address some of the inherent difficulties in evaluating measures of biodiversity conservation.

4. BIODIVERSITY CONSERVATION INDEX

As pointed out earlier, biodiversity conservation is an extraordinarily broad concept. Because of this, traditional approaches to assessing biodiversity (e.g., use of 'keystone' species, use of indices such as abundance, richness, evenness, or the Shannon Index) are not adopted. Instead, assessment of biodiversity conservation is geared towards a general examination of the management practices affecting biodiversity and the state or condition of the processes that generate or maintain biodiversity. This approach is consistent with the C&I methodology for assessing sustainable forest management as discussed in Mendoza and Prabhu (1999).

The paper follows the AHP framework described in Mendoza and Prabhu (1999). Hence, the basic structure also is hierarchical where the different indicators described in Boyle *et al.* (1996) are organised at different levels as shown in Figure 1. Assessment is made at each level. From the figure, it can be seen that the Biodiversity Conservation Index can be estimated in a two-stage process consistent with a two level hierarchy. The first level can be viewed as 'indicators' while the second level corresponds to the 'verifiers'. For the purpose of this paper, the two lower level conceptual tools used in C&I assessment are briefly defined below.

Indicator: a variable or component of the forest or the relevant management system used to infer attributes of the sustainability of the resource and its utilisation

Verifier: a set of values that define suitable reference conditions for an indicator.

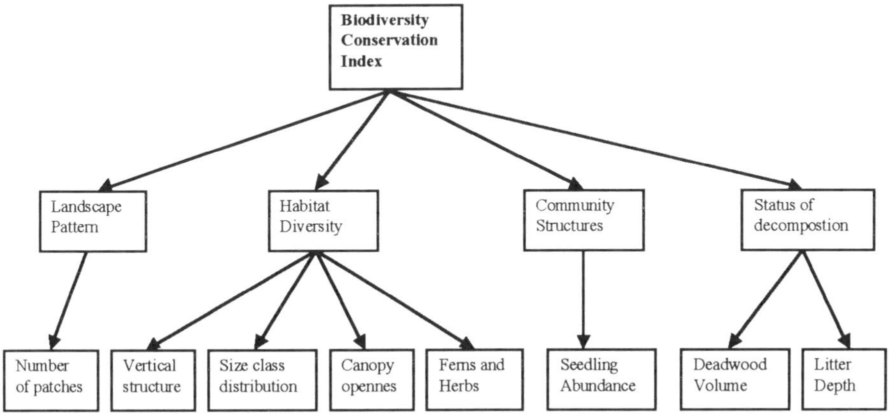

Figure 1. C&I hierarchy for biodiversity conservation.

5. FIRST LEVEL ANALYSIS

Following the terminology described above, the first-level analysis is conducted at the indicator level. That is, biodiversity conservation is measured based on a number of indicators. From the definition above, these indicators are not directly measurable themselves, but are represented by the cumulative attributes of the verifiers. These verifiers provide specific details about the indicator, and therefore constitute the primary source of information or data for analysing biodiversity conservation.

From Figure 1, it is clear that biodiversity conservation must be assessed as a composite measure reflecting the cumulative effects of all indicators. Hence, the impacts of all indicators must be aggregated. In this paper, a simple method of aggregation involving the 'linear combination' of all indicators is used. This method was chosen because of its simplicity and transparency where the cumulative effect is aggregated by simply adding the individual effects of all indicators.

The process of aggregation brings another issue to the assessment. Invariably, some indicators can be viewed as relatively more significant than others. Hence, their impacts must be accorded more importance compared to other less significant indicators. These degrees of importance must be reflected in the overall composite measure achieved through aggregation. In light of this, each indicator must be assigned a measure of relative importance.

In like manner, each indicator also has a number of verifiers whose impact to the indicator has varying degrees of importance. Therefore, aggregating the effects of all verifiers also must consider the relative importance of each verifier.

Based on the above discussion, the first-level analysis involves estimating the cumulative impacts of verifiers at the indicator level that can now be formulated as:

$$I_j = \sum w_x \mu_x \tag{5.1}$$

where w_x represents the relative importance of verifier x, and μ_x is the measure of sustainability associated with verifier x. The relative weights, w_x, are calculated based on the pairwise comparisons of the verifiers following the AHP procedure described in Mendoza and Prabhu (1999) and explained in more detail by Saaty (1996). These relative weights are normalised and scaled such that, $0 \leq w_x \leq 1$, and $\Sigma w_x = 1$. As described in the next section, μ_x is also parameterised such that, $0 \leq \mu_x \leq 1$.

I_j is an index that provides a rough indication of the cumulative impacts of all verifiers on the favourability of indicator j to biodiversity conservation. Since both w_x and μ_x are normalised and scaled, I_j is also normalised and scaled such that $0 \leq I_j \leq 1$. As such, the index itself can be interpreted as follows. A high I_j value implies that the indicator is favourable to biodiversity conservation; low I_j value implies that the indicator contributes poorly to biodiversity conservation. Values between 0 and 1 reflect varying degrees of favourability to biodiversity conservation.

6. A FUZZY METHOD TO EVALUATE VERIFIERS

As stated earlier, verifiers are the measurable components of the C&I approach to forest sustainability analysis. Similarly, the verifiers constitute most of the observable and measurable attributes of biodiversity conservation analysis. While these verifiers may be apparent and lend themselves to direct measurement, it may be quite difficult to judge their effects accurately. In the context of biodiversity conservation, for example, some verifier attributes may be known to have some impact; however, the extent or magnitude of its impact may be difficult or impossible to evaluate. In other words, there is some uncertainty associated to the evaluation itself; an uncertainty that is not due to the classical case of randomness. Rather, the uncertainty is engendered by the intrinsic complexity or ambiguity associated with the verifier itself and its relationship to the indicator. In view of this, classical evaluation methods based on 'crisp' measurement philosophy cannot be used. For instance, there may be verifiers depending on the assessment of their values for which it can not be ascertained whether they lead to favourable conservation of biodiversity. In other words, the impact of the verifiers can only be judged in terms of the degree to which they lead to favourable or unfavourable conservation of biodiversity. In this context, classical assessment methods that directly assign whether the verifier is favourable or not, can not be used. Instead, a fuzzy method of assessment based on fuzzy set theory (Zadeh 1965) is adopted. The next section describes some of the principles of the method briefly. For detailed description of the method and the theory behind it, readers are referred to other published materials (e.g., Zimmerman 1985).

Zadeh (1965) developed the concept of fuzzy sets as a basis for approximate reasoning, and to accommodate imprecision and uncertainties. The fundamental concept of fuzzy sets, one that has direct relevance to the sustainability assessment process, is the "*membership function.*" The premise of fuzzy logic is that "*membership*" to a set is not dichotomous (i.e., in or out, true or false); instead, there are degrees of membership ranging between 0 to 1. This theoretical construct has a direct parallel to the concept of sustainability. Since it is highly unlikely that precise estimates can be made on the sustainability of forests, it is more meaningful to characterise assessments in terms of degrees of sustainability. Hence, forests with degrees of membership close to one imply "close to being sustainable" and vice versa.

Following the fuzzy set concept and the membership function described above one can define sustainability as a fuzzy set where the membership function can be defined as follows:

$$\mu_x = \begin{cases} 0 & \text{if } x \le \alpha \\ 1 - \dfrac{\beta - x}{\beta - \alpha} & \text{if } \alpha < x < \beta \\ 1 & \text{if } x \ge \beta \end{cases} \tag{6.1}$$

where α and β are parameters representing limits or threshold values with regards to sustainability; x is the value of the verifier.

The above formulation is a linear membership function (Zimmerman 1985). Other more complex forms of membership functions may be used. For example, Mendoza and Prabhu (1998) describe other forms of membership functions representing different types of possibility configurations.

7. SECOND LEVEL ANALYSIS: THE BIODIVERSITY CONSERVATION INDEX (BCI)

After aggregating the impacts of verifiers for each indicator, the next level in the hierarchy is the analysis of all indicators and their cumulative impacts on biodiversity conservation. In other words, this analysis involves the assessment of biodiversity conservation index itself. The process entails the aggregation of the favourability measures of each indicator as estimated in level 1. Hence, the BCI can be formulated as:

$$BCI = \sum s_j I_j \tag{7.1}$$

where s_j are the relative weights of indicator j, such that $\Sigma s_j = 1$; and I_j is its corresponding favourability index or value as estimated in (5.1). Like the analysis at the indicator level where verifiers are assigned different measures of relative importance, the indicators under biodiversity conservation must also be given different measures of importance depending on their perceived significance or impact. This relative importance, denoted by s_j is calculated by AHP using the pairwise comparisons of all indicators.

8. BIODIVERSITY CONSERVATION CASE STUDY

A simple case study is presented to illustrate the application of the models described earlier. The data set was obtained from a logging concession located in central Kalimantan, Indonesia. Twelve sample plots were established at strategic locations within the site. At each plot, data were collected to represent most of the verifiers described in Figure 1. Summary statistics of each verifier are contained in Table 1.

To conduct the biodiversity conservation study, four biodiversity experts were selected. All are scientists based at the Centre for International Forestry Research (CIFOR) located at Bogor, Indonesia. Their primary roles were to make pairwise comparisons of the different elements of biodiversity conservation; both at the indicator level, and at the verifier level. Questionnaires and response forms were distributed to each expert. Prior to soliciting their expert opinions and judgement, the experts were convened for the purpose of explaining the AHP methodology and the context with which it is applied. All of them were familiar with the principles of C&I; they were also well informed and knowledgeable about the indicators and verifiers of biodiversity conservation described in Boyle *et al.* (1997). At the meeting, the experts were allowed to exchange their views about the elements of the biodiversity conservation hierarchy. However, each expert responded to the questionnaire individually.

Table 1. Summary statistics of verifier data obtained from 12 sample plots.

Plot	Number of Patches[a]	Vertical Structure (m)	Size Classes (cm)	Canopy Openness (%)	Herbs[b]	Abundance of Seedlings	Depth of litter (cm)	Deadwood volume
1	2	21.02	23.29	78.83	2	380	3.2	19.21
2	1	24.66	21.50	84.58	3	560	3.4	24.21
3	3	22.69	21.49	87.95	5	601	4.1	23.73
4	2	20.19	23.43	82.21	3	568	2.2	17.07
5	1	18.97	21.68	73.71	4	467	4.6	35.42
6	1	21.11	22.81	72.75	6	447	5.4	22.43
7	2	20.86	23.89	79.67	3	567	5.6	33.59
8	2	16.69	17.46	83.54	4	678	2.5	18.82
9	2	22.11	24.99	80.29	6	326	3.6	16.90
10	1	17.33	22.11	89.17	4	435	4.3	31.46
11	2	18.53	54.38	85.13	4	478	3.8	35.32
12	2	14.67	16.02	80.79	6	562	3.8	55.07

[a]These were hypothetically estimated for illustrative purposes.
[b]This is plant cover with categorical data as follows: 0 = 0 %, 1 = scarce, 2 = scattered, 3 = scattered, 4 = 5 %, 5 = 20 %, 6 = 25 - 33 %, 7 = 33 - 50 %, 8 = 50 - 75 %, 9 = > 75 %, 10 = 100 %.

The biodiversity conservation hierarchy adopted in the study is shown in Figure 1. Note that the verifiers used are only a subset of those generated for the area by Boyle *et al.* (1997). The subset of verifiers was chosen mainly because of data availability.

9. ANALYSIS OF RESULTS

The case study used four indicators and eight verifiers. As described earlier, the first step in the AHP process is the decomposition of the problem into a hierarchy of elements. This is illustrated in Figure 1 where, biodiversity conservation has four indicators and eight verifiers.

The second step is the assessment of the elements of the hierarchy. This was done at two levels: at the indicator level and the verifier level. At both levels, the experts were asked to conduct pairwise comparisons of all indicators and verifiers within each indicator.

9.1 Analysis at the Indicator Level

Table 2 summarises the results of the indicator analysis. The table contains the relative weights estimated from the pairwise comparisons.

Three experts had to perform the pairwise comparisons twice before an acceptable level of inconsistency denoted by the inconsistency index (ICI) was achieved. One of the four experts (see column 3 in Table 2) generated an acceptable set of pairwise comparisons (i.e. ICI less than 10%) after one iteration.

Table 2. Relative importance of indicators in percent

	Expert Evaluations								
Indicator	1		2		3		4		Average
Iteration[a]	1	2	1	2	1	2	1	2	
Landscape Pattern	5	5	10	10	56	56	53	23	28
Change in Diversity	14	14	25	12	27	27	21	57	28
Community Structures	22	26	38	38	6	6	14	13	24
Status of Decomposition	59	55	27	40	11	11	12	7	21
Inconsistency Index	13	9	15	1	3	3	17	5	

[a]Iteration number denotes the number of iterations the expert performed the pairwise comparisons before the Inconsistency Index was below 10%. The "Iteration 2" columns denote the relative weights based on improved pairwise comparisons (no higher than 10%) which was used to determine the average weights for all indicators.

The concept of inconsistency is quite useful in the context of multicriteria analysis. In the AHP, the inconsistency index is a measure of the logical (in)consistency of the experts' judgements based on their pairwise comparisons. It provides consistency information reflecting both the ordinal and cardinal importance of the two elements compared. In general, a tolerance (in)consistency index of 10% is acceptable for comparisons involving no more than 9 elements (Saaty, 1995). Higher inconsistency levels may be tolerable for comparisons involving more than 9 elements.

From Table 2, only expert 3 generated a highly consistent assessment in the first iteration. The other three experts had to conduct a second round of pairwise comparisons before a consistent set of judgements was achieved. Before the second iteration, the three experts were informed that the AHP is capable of 'guiding' their assessments to arrive at an improved (i.e. lower inconsistency index) set of comparisons following the method of Saaty (1995). The three declined to use such guidance because of their concern that it may bias their assessments. The second round assessments all yielded more consistent comparisons (i.e. all were below 10% inconsistency).

9.2 Analysis at the Verifier Level

Table 3 contains the results of the AHP analysis on the verifiers. Because only Indicator 2 has more than two verifiers, only its verifiers were subjected to AHP analysis. Again, all experts were asked to make their judgements via pairwise comparisons of all four verifiers of Indicator 2. Experts 2 and 3 both needed only one iteration to arrive at a consistent set of comparisons (i.e. ICI less than 10%). Experts 1 and 4 had to do the comparisons twice before arriving at acceptable comparisons. Table 3 contains the relative weights of all verifiers based on the pairwise comparisons. The average weights were calculated based on the improved relative weights that were calculated from the pairwise comparisons with lower than 10% inconsistency.

Table 3. Relative importance of verifiers in percent

Indicator 2	Expert Evaluations								
Verifiers	1		2		3		4		Average
Iteration[a]	1	2	1	2	1	2	1	2	
Vertical Structure	25	28	61	61	58	58	33	16	41
Size Class	9	10	18	18	25	25	28	46	26
Canopy Openness	61	57	12	12	10	10	19	23	25
Herbs	5	6	9	9	7	7	20	15	9
ICI	25	7	8	8	5	5	15	10	
Indicator 4									
Deadwood volume		22		50		64		50	47
Depth of Litter		78		50		36		50	53

[a]Iteration number denotes the number of iterations the expert performed the pairwise comparisons before the Inconsistency Index was below 10%. The iteration 2 columns denote the relative weights based on improved pairwise comparisons (no higher than 10%) which was used to determine the average weights for all indicators.

9.3 Fuzzy Evaluation of Verifiers

Before the biodiversity conservation index can be calculated for the case study area, the verifiers must be evaluated in terms of how favourable they are to biodiversity conservation. Recall that this type of assessment is handled using fuzzy methods as shown by the formula in (6.1). In this study, the simple linear membership function is adopted.

Table 4 contains the information necessary to define the membership function of the verifiers. Figure 2 graphically describes the linear membership function given the parameters α and β. The lowest and highest values are the 'observed' values. On the other hand, the α and β entries are limits of the membership function as shown the formula in (6.1).

Table 4. Membership functions of fuzzy verifiers

Verifiers	Lowest Value	Highest Value	Linear α	β
Number of Patches	1	4	0	4
Vertical Structure	15	27	5	20
Size Classes	2	37	5	20
Canopy Openness	68	90	50	80
Herbs[a]	0	10	1	10
Abundance of Seedlings	340	704	360	600
Deadwood Volume	13.153	55.761	10	40
Depth of Litter	2	6	0	10

[a]Herbs is classified as a categorical data between 0 – 10; each category reflecting the amount of herbs (e.g., scarce, scattered, 5% of plant cover, 20% of plant cover, etc.)

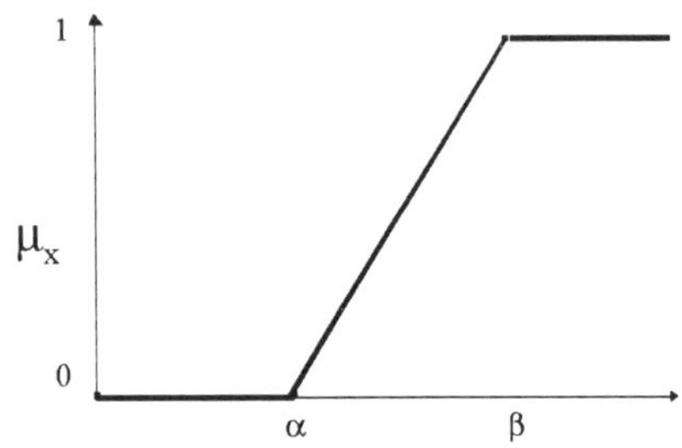

Figure 2. Membership function of fuzzy verifiers.

9.4 Biodiversity Conservation Index

One of the main objectives of this paper is to develop a biodiversity conservation index that could be used as a criterion for assessing forest sustainability. The hierarchical model described in Figure 1, and their corresponding functional models formulated in the formulae (5.1, 6.1, and 7.1) allow the estimation of such a biodiversity conservation index. Note that the indices were calculated in a two-steps process following the two-stage or two-level analysis described earlier. The first-level analysis involves the analysis at the indicator level; the second analysis is done at the verifier level.

At the first level, the verifiers of each indicator were analysed resulting in the estimation of their relative weights (Table 3). These values correspond to the relative weights w_x in the formula in (5.1). In addition, the membership values of each verifier were also calculated using the parameters α and β (Table 4). These membership values correspond to the μ values contained in Table 5 and modelled as μ_x in the formulae (5.1) and

(6.1). Given the relative weight w_x and the membership values μ_x of the verifiers, the indicator favourability values, I_j are estimated (Table 6). With the relative importance values of each indicator contained in Table 2 (corresponding to s_j in the formula in (7.1)) the biodiversity conservation index can be calculated using the formula in (7.1) as shown in Table 6.

Table 5. Membership function values of each plot based on mean values of verifiers.

Plot		Number of Patches	Vertical Structure (m)	Size Classes (cm)	Canopy Openness (%)	Herbs	Abundance of Seedlings	Depth of litter (cm)	Deadwood volume
1	Mean	2	21.02	23.29	78.83	2	380	3.2	19.21
	μ	0.5	1	1	0.96	0.11	0.60	0.32	0.30
2	Mean	1	24.66	21.50	84.58	3	560	3.4	24.21
	μ	0.25	1	1	1	0.22	0.92	0.34	0.47
3	Mean	3	22.69	21.49	87.95	5	601	4.1	23.73
	μ	0.75	1	1	1	0.44	1	0.41	0.45
4	Mean	2	20.19	23.43	82.21	3	568	2.2	17.07
	μ	0.5	1	1	1	0.22	0.94	0.22	0.23
5	Mean	1	18.97	21.68	73.71	4	467	4.6	35.42
	μ	0.75	0.93	1	0.79	0.33	0.76	0.46	0.84
6	Mean	1	21.11	22.81	72.75	6	447	5.4	22.43
	μ	0.5	1	1	0.75	0.55	0.72	0.54	0.41
7	Mean	2	20.86	23.89	79.67	3	567	5.6	33.59
	μ	0.25	1	1	0.98	0.22	0.94	0.56	0.78
8	Mean	2	16.69	17.46	83.54	4	678	2.5	18.82
	μ	0.75	0.77	0.83	1	0.33	1.13	0.25	0.29
9	Mean	2	22.11	24.99	80.29	6	326	3.6	16.90
	μ	0.5	1	1	1	0.55	0.51	0.36	0.23
10	Mean	1	17.33	22.11	89.17	4	435	4.3	31.46
	μ	0.75	0.82	1	1	0.33	0.70	0.43	0.71
11	Mean	2	18.53	54.38	85.13	4	478	3.8	35.32
	μ	0.25	0.90	1	1	0.33	0.78	0.38	0.84
12	Mean	2	14.67	16.02	80.79	6	562	3.8	55.07
	μ	0.75	0.64	0.73	1	0.55	0.93	0.38	1

10. CONCLUSIONS

The information contained in Table 6 is the result of a number of assessments, and hence can be viewed as reflecting the cumulative effects of all assessments. Firstly, consider the assessments of all verifiers and indicators. Because biodiversity conservation is a broad concept, it was useful and meaningful to deconstruct the concept into levels of more manageable elements (i.e., indicators and verifiers) and organise them into a hierarchy. Such an approach allows the individual assessments of verifiers and their collective impacts on the indicators. This structured analysis enables a 'tractable' way of doing the analysis; thereby offering justification for the results of the assessment.

Table 6. Indicator favourability values (I_j) and the biodiversity conservation index of plots.

Plot	Landscape Pattern	Change in Diversity	Community Structures	Status of Decomposition	Biodiversity Conservation Index
1	0.5	0.92	0.60	0.324	0.60964
2	0.25	0.94	0.92	0.37	0.6317
3	0.75	0.96	1	0.42	0.807
4	0.5	0.94	0.94	0.22	0.675
5	0.75	0.86	0.76	0.54	0.7466
6	0.5	0.91	0.72	0.51	0.6747
7	0.25	0.94	0.94	0.61	0.6869
8	0.75	0.81	1.13	0.26	0.7626
9	0.5	0.97	0.51	0.33	0.6033
10	0.75	0.87	0.70	0.49	0.7245
11	0.25	0.91	0.78	0.48	0.6128
12	0.75	0.75	0.93	0.67	0.7839

Secondly, the verifiers and indicators are assessed relative to their perceived degree of importance. Hence, the more important individual elements are, the more it impacts higher level analysis. This differentiated analysis provides a more objective evaluation of all verifiers and indicators.

Thirdly, the assessments were based on the evaluation of all experts; hence, it is a group evaluation rather than the result of biased opinions of one or selected group of experts and stakeholders. This group decision-making feature of the model promotes participative and collective involvement of various groups, and enhances the acceptability of any assessment made.

Fourthly, the final result of the assessment is not a judgement of whether forests are sustainable or not; rather, the index reflects degrees of sustainability. Hence, the higher the value of the index, and the closer it is to the value 1, the more sustainable the forest is judged to be. The opposite is also true; the lower the value of the index, the forest is judged less sustainable.

In addition to the information generated from the biodiversity index, the models described here, and the other information in Tables 5 and 6, could also be used as general guides to forest managers. For example, the favourability values of indicators are one piece of information that managers could use to target forest activities to those indicators that have low I_j values in order to increase the likelihood that the forest can be managed sustainably.

This paper described how AHP could be used as a formal analytical framework with which biodiversity conservation can be assessed. Experience gained from the case study showed that the model is very applicable; it has desirable properties that make it a powerful tool to conduct a broad forest sustainability assessment such as the biodiversity conservation study described in this paper.

REFERENCES

BOYLE, T., LAWES, M., MANOKARAN, N., PRABHU, R., GHAZOUL, J., SASTRAPRADJA, S., THANG, H., DALE, V., EEELEY, H., FINEGAN, B., SOBERON, J. AND STORK, N. 1997. Criteria and Indicators for Assessing the Sustainability of Forest Management. A Practical Approach to Assessment of Biodiversity. Center for International Forestry Research (CIFOR). CIFOR Special Paper. Bogor, Indonesia.

MENDOZA, G.A. AND PRABHU, R. 2000. Prioritising Criteria and Indicators for Sustainable Forest Management: A Case Study on Participatory Decision Making. In The Analytic Hierarchy Process for Natural Resources and Environmental Decision-Making. Schmoldt, D., Kangas, J., Mendoza, G., and Pesonen, M. (eds.). Kluwer Academic Publishers, Amsterdam.

MENDOZA, G.A. AND PRABHU, R. 1999. Integrating Multiple Criteria Decision Making and Geographic Information Systems for Assessing Indicators of Biodiversity Conservation. Working Paper GAM-99-1. University of Illinois, Department of Natural Resources and Environmental Sciences. Urbana, Illinois.

SAATY, T. 1995. Decision making for leaders: The Analytic Hierarchy Process in a Complex World. RWS Publications.

SAATY, T. 1996. MultiCriteria Decision Making: The Analytic Hierarchy Process. University of Pittsburgh.

STORK, N., BOYLE, T., DALE, V., EELEY, H., FINEGAN, B., LAWES, M., MANOKARAN, M., PRABHU, R. AND SABERON, J. 1997. Criteria and Indicators for Assessing the Sustainability of Forest Management: Conservation of Biodivesity. Occasional Paper No. 17. Centre for International Forestry Research Bogor, Indonesia.

ZADEH, L. 1965. Fuzzy sets. Inf. Control 8:338-353.

ZIMMERMANN, H.J. 1985. Fuzzy set theory and application. Nijhoff Publishing, Boston, MA

Chapter 15

Regression Methods for Pairwise Comparison Data

Juha M. Alho and Osmo Kolehmainen
University of Joensuu, Department of Statistics, Joensuu Finland

Pekka Leskinen
Finnish Forest Research Institute, Kannus Research Station, Kannus Finland

Key words: Bayes, experimental design, explanatory variables, statistical inference, uncertainty, variance components

Abstract: Multi-objective decision making often requires the comparison of qualitatively different entities. For example, a forest owner has to assess the aesthetic and recreation values of the forest in addition to the income from selling wood. Pairwise comparisons can be used to elicit relative preferences concerning such entities. Eigenvalue techniques introduced by Saaty (1977) are one way to analyse pairwise comparisons data. A weak point of the original methodology has been that it does not allow a statistical analysis of uncertainties in judgements. The eigenvalue technique also requires that all entities have been compared with each other. In many applications, this is impracticable because of the large number of pairs. The number of judges can also be large, and there can be missing observations. Moreover, it is frequently of interest to analyse how different attributes of the entities, or different attributes of the judges, influence the relative preference. In this paper, we first review our previous work with an alternative methodology based on regression analysis. Then, we show how explanatory variables can be incorporated. The construction of the design matrix is detailed and the interpretation of the results is discussed.

1. INTRODUCTION

Saaty (1977) introduced the so-called analytic hierarchy process (AHP) as a method of deriving a ratio scale of preferences (or priorities) concerning a set of *m* entities or attributes. The method involves a quantification of all

D.L. Schmoldt et al. (eds.),
The Analytic Hierarchy Process in Natural Resource and Environmental Decision Making, 235–251.

$m(m-1)/2$ pairwise comparisons between the entities. The ratio scale is derived using an eigenvalue calculation on a matrix formed from the quantified comparisons. De Jong (1984) and Crawford and Williams (1985) showed how regression techniques could be used to provide alternative estimates. Later, the regression approach has been used by Carriere and Finster (1992) and Zhang and Genest (1996), for example; related probabilistic formulations have been provided by Basak (1989, 1990, 1991). Typically the two methods give similar numerical results. Empirical and theoretical evidence for this has been provided by Budescu, Zwick and Rapoport (1986), Zahedi (1986) and Genest and Rivest (1994). However, if the comparisons are severely inconsistent, the results may differ considerably (Saaty and Vargas 1984).

In Alho, Kangas and Kolehmainen (1996) we extended the regression approach to the case of multiple judges, and introduced a variance components model for the analysis of the inconsistency of the evaluations. In Alho and Kangas (1997) we provided a Bayesian formulation of the regression approach. This work has been further developed by Leskinen and Kangas (1998), who considered pairwise comparisons data elicited in a sequence of two questions. The first asks for a relative preference, and the second for a subjective quantification of the uncertainty of the first response.

In this paper we review our previous work on the regression approach. As a new development we show how the characteristics of the entities being compared (landscape pictures, in our illustrations), or the background characteristics of the judges, can be used to model preferences via regression, in practice. The difficulty is with the creation of a non-standard design matrix. If the background characteristics are categorical (e.g., a landscape might be evaluated by local people, tourists, or expert ecologists), then by stratifying the data one could, in principle, use the eigenvalue calculations by stratum. However, this is clumsy whenever the number of strata is large. Moreover, if the background characteristics are continuous (such as age), then the approach may be completely infeasible. The theory presented is closely related to the analysis of pairwise comparisons in which the probability that one attribute be preferred over another is estimated in a study population (see Dittrich *et al.* 1998). The theory is also related to the choice models of McFadden (1974, 1981) that have been used to analyse consumer behaviour and geographic mobility, for example.

Another motive is to illustrate how the regression approach permits the estimation of the relative scale based on fewer comparisons. This has been suggested earlier by Carriere and Finster (1992), for example. The minimum is one less than the number of entities. Of course, with the minimum number of comparisons the quality of the estimates would be expected to be poor. Therefore, deciding on an intermediate value becomes an important part of the experimental design.

Section 2 reviews the regression approach. In Section 3 we will formulate a loglinear model with explanatory variables, and derive some implications for the resulting ratio scales. Section 4 derives the design matrix for pairwise comparisons data under the framework of Section 3, and Section 5 discusses the planning of pairwise experiments. An application to landscape evaluations in discussed in Section 6, and Section 7 discusses the implications of the regression methodology. Section 2.2 and Chapter 4 are somewhat technical in nature and can be omitted at first reading.

2. REGRESSION APPROACH

2.1 Model Formulation

Alho *et al.* (1996) studied the uncertainty in expert predictions concerning the ecological consequences of 10 alternative forest plans with respect to the habitat requirements of black grouse, a valued game-bird. The relative merits of the plans were evaluated in a pairwise manner by 15 experts. All $10 \times 9 / 2 = 45$ pairwise comparisons were made by each of the experts. We show how regression analysis can be used to analyse such data.

Let v_i be the value of entity (e.g. forest plan) $i = 1,\ldots,I$, and let $r_{ii'k}$ be the ratio $v_i / v_{i'}$ as perceived by judge $k = 1,\ldots,K$. Saaty (1977) suggested that scores 1/9, 1/8, ..., 1/2, 1/1, 2/1, ..., 8/1, 9/1 be used in the elicitation of the ratios $r_{ii'k}$. Alternative scores have been proposed by Lootsma (1993), and Salo and Hämäläinen (1997), for example.

Because the v_i are positive, it can assumed without loss of generality that $v_i = \exp(\mu + \alpha_i)$, so the theoretical value of the ratio $v_i / v_{i'}$ is $\exp(\alpha_i - \alpha_{i'})$. However, due the difficulty of giving consistent evaluations of all the pairs, we expect there to be deviations from the theoretical value. For example, if we prefer i to i' by 2 to 1, and i' to i'' by 3 to 1, then we should prefer i to i'' by 6 to 1. If we don't, then we are inconsistent. In general, the pairwise comparisons of judge k are consistent, if $r_{ii''k} = r_{ii'k}\, r_{i'i''k}$ for every i, i', and i'' (Saaty 1977). Other sources of inconsistency are the problems of numerical scaling, and disagreements between judges.

Define $y_{ii'k} = \log(r_{ii'k})$. Then the regression model for pairwise comparisons data in the multiple judge case is (Alho *et al.* 1996)

$$y_{ii'k} = \alpha_i - \alpha_{i'} + \varepsilon_{ii'k} \tag{2.1}$$

where the error term representing all types of inconsistencies has expected value $E[\varepsilon_{ii'k}] = 0$. For identifiability, it is assumed that $\alpha_I = 0$, so α_i measures the value of entity i relative to entity I.

2.2 Least Squares Estimation

Take $I = 4$ for example, and define a vector $\mathbf{Y}$ which consists of subvectors $\mathbf{Y}_k$, $k = 1,\ldots,K$ where $\mathbf{Y}_k = (y_{12k}, y_{13k}, y_{14k}, y_{23k}, y_{24k}, y_{34k})^{\mathrm{T}}$. Define also an error vector $\boldsymbol{\varepsilon}$ consisting of subvectors $\boldsymbol{\varepsilon}_k$ analogous to those of $\mathbf{Y}$, and let $\alpha = (\alpha_1, \alpha_2, \alpha_3)^{\mathrm{T}}$. Furthermore, let $\mathbf{X} = \mathbf{1}\otimes\mathbf{M}$, where $\mathbf{1}$ is a K-vector of 1's, the symbol $\otimes$ represents the *Kronecker product*, and

$$\mathbf{M} = \begin{bmatrix} 1 & -1 & 0 \\ 1 & 0 & -1 \\ 1 & 0 & 0 \\ 0 & 1 & -1 \\ 0 & 1 & 0 \\ 0 & 0 & 1 \end{bmatrix} \tag{2.2}$$

Then the regression model (2.1) gets the form $\mathbf{Y} = \mathbf{X}\boldsymbol{\alpha} + \boldsymbol{\varepsilon}$, and the ordinary least squares (OLS) solution for the vector $\boldsymbol{\alpha}$ is $\hat{\boldsymbol{\alpha}} = (\mathbf{X}^{\mathrm{T}}\mathbf{X})^{-1}\mathbf{X}^{\mathrm{T}}\mathbf{Y}$ (Alho *et al*. 1996). Estimates $\hat{\alpha}_i$ can be transformed to the scale of the priorities by $\exp(\hat{\alpha}_i)/\Sigma_j\exp(\hat{\alpha}_j)$, where $\hat{\alpha}_I = 0$. The form of $\mathbf{M}$ for general I follows the same pattern as (2.2).

Note that $\mathbf{1}\otimes\mathbf{M}$ simply means that K matrices $\mathbf{M}$ have been stacked, so in a case of single judge we have that $\mathbf{X} = \mathbf{M}$.

Each row of $\mathbf{M}$ corresponds to a pairwise comparison and each column to a parameter to be estimated. In the single judge case the equation (2.1) becomes $y_{ii'} = \alpha_i - \alpha_{i'} + \varepsilon_{ii'}$, where the error terms are uncorrelated with mean zero and $\mathrm{Var}(\varepsilon_{ii'}) = \sigma^2$. An unbiased estimator for the residual variance is then $\hat{\sigma}^2 = (\mathbf{Y} - \mathbf{X}\hat{\boldsymbol{\alpha}})^{\mathrm{T}}(\mathbf{Y} - \mathbf{X}\hat{\boldsymbol{\alpha}})/(n - I + 1)$, where $n = I(I-1)/2$. Residual variance is a natural measure of inconsistency.

2.3 Variance Components

In the multiple judge case (and under certain restricted models) the assumption of independence for the error terms may not be realistic. One possible dependency structure for the error term $\varepsilon_{ii'k}$ of model (2.1) can be given as follows (Alho *et al*. 1996). First, suppose the actual value of entity i for judge k deviates from the population mean so that it can be written as

$\exp(\mu+\alpha_i+\eta_{ik})$. Second, suppose there is a bias specific to each pair (i, i') that is shared by all judges, so the relative value of i to i' for judge k is of the form $\exp(\alpha_i-\alpha_{i'}+\eta_{ik}-\eta_{i'k}+\xi_{ii'})$. Third, adding a term for residual error we get that $r_{ii'k}=\exp(\alpha_i-\alpha_{i'}+\eta_{ik}-\eta_{i'k}+\xi_{ii'}+\delta_{ii'k})$. This leads to the variance components representation

$$\varepsilon_{ii'k} = \eta_{ik} - \eta_{i'k} + \xi_{ii'} + \delta_{ii'k} \tag{2.3}$$

where the expectation of each term on the right hand side is zero. We assume that the random effects are independent with $\mathrm{Var}(\delta_{ii'k}) = \sigma_1^2$, $\mathrm{Var}(\xi_{ii'}) = \sigma_2^2$, and $\mathrm{Var}(\eta_{ik}) = \sigma_3^2$. Now the interindividual variation is represented by σ_3^2, the inconsistency shared by all judges is represented by σ_2^2, and residual inconsistency specific to a judge is represented by σ_1^2.

Methods for the estimation of the variance components as well as alternatives to OLS estimation for $\boldsymbol{\alpha}$ can be found in Alho, Kangas and Kolehmainen (1996). The methods were applied to a data set concerning the ecological consequences of forest plans that was mentioned in the beginning of Section 2.1.

Lack of consistency may sometimes be decreased by the iterative Delphi-technique (cf., Linstone and Turoff 1975). In this approach a group of experts are interviewed, the elicited views are shared, and the experts are re-interviewed. The sequence may be repeated to see if there is a convergence of views. An application of the variance components formulation in this setting is presented in Kangas, Alho, Kolehmainen and Mononen (1998), where the object is to assess alternative forest plans from the point of view of biodiversity. Changes in the size of the variance components were used to quantify the success of the Delphi rounds. Overall, the Delphi rounds decreased inconsistency, but the decline was not uniform for all variance components.

2.4 Bayesian Analysis

Alho and Kangas (1997) studied the problem of choosing an optimal forest plan for a state-owned forest area of 321 ha in Kuusamo, North-Eastern Finland. The highest level of the decision hierarchy consisted of three sources of utility: timber production, scenic beauty, and game management. Their relative priorities were evaluated by the staff of the Finnish Forest Park Service (FPS) using pairwise comparisons. Each of these was further decomposed into two or three sub-criteria. The priorities of the sub-criteria within each source of utility were similarly quantified by the FPS staff. Finally, six different plans (continue natural growth with no cuttings; optimise scenic beauty index; normal forestry guidelines; optimise game values; modest regeneration; maximise income) were evaluated using pairwise comparisons with respect to each of the sub-criteria. Comparisons between forest plans with respect to timber production and scenic beauty were made by the timber planning staff of the FPS, and with respect to game habitat by the game management experts of the FPS. The goal was to quantify the priorities and their uncertainty so that it would best represent the interests of the public at large. A Bayesian formulation seemed to provide a natural language for the task.

Each round of pairwise comparisons was carried out by a group acting as a single judge, so we had $K = 1$, for every regression. In the decision hierarchy, the model (2.1) was applied repeatedly to quantify the priorities of the three utilities, their sub-criteria, and the forest plans with respect to the sub-criteria (Alho and Kangas 1997). A Bayesian interpretation of the results was obtained by assuming non-informative prior proportional to σ^{-1} for the pair $(\boldsymbol{\alpha}, \sigma)$. The resulting posterior distribution of $\boldsymbol{\alpha}$ is well-known (cf., Box and Tiao 1973, 117). The posterior distributions for the priorities were estimated numerically by simulation by taking samples from the posterior of $\boldsymbol{\alpha}$ (Alho and Kangas 1997). This allowed the computation of probabilities for events like "forest plan A is better than B," for example. The Bayesian approach had two advantages over the frequentist formulation. First, since the priorities are a non-linear function of the parameters $\boldsymbol{\alpha}$, the derivation of the second moments of the priority estimators (via the delta-method) is technically complicated, and potentially inaccurate because of the small sample size. Second, the Bayesian posterior probabilities may be more easily understood by decision-makers than *p*-values provided by the frequentist analysis.

2.5 Interval Judgements

When pairwise comparisons are perfectly consistent, then $\hat{\sigma}^2 = 0$ in the single judge model, otherwise $\hat{\sigma}^2 > 0$. Since only a small number of scores are used to quantify the elicited preferences, the residual variance may be zero although there is genuine uncertainty about the priorities.

To capture this type of uncertainty, Leskinen and Kangas (1998) suggested that one use interval judgements instead of judgements given as a single number. They used the Bayesian regression framework of Section 2.4, and applied the method to the same decision hierarchy concerning state owned forest in Kuusamo that was described in Section 2.4. The interval judgements were elicited in a sequence of two questions. First, the ratio $r_{ii'}$ was elicited. Then, an upper and a lower limit around the elicited $r_{ii'}$ were defined, and the judge was asked to evaluate the probability that his or her true preference lies between the limits. By assuming an underlying normal distribution for the true preferences, a direct specification of the probability distribution of the response **Y** becomes available that represents both the preferences and their uncertainty. Samples can be generated from the distribution and corresponding regression estimates calculated. This yields a posterior distribution of the regression parameters (Leskinen and Kangas 1998). The procedure leads to a positive posterior variance also when single pairwise comparisons are consistent.

3. A MODEL WITH EXPLANATORY VARIABLES

The methods reviewed above have primarily been developed to give a statistical assessment of the uncertainty in the elicited priorities. In many cases, there is interest in the understanding of the factors that might explain variations in the preferences across judges, or across entities being compared.

Suppose we have judges $k = 1, \ldots, K$ whose task is to evaluate photographs taken at locations $i = 1, \ldots, I$. A treatment $j = 1, \ldots, J$ is applied to each photograph to reflect possible future use of the landscape depicted. In our examples, this was done by editing a digitised photograph by computer. We will speak of photograph "(i, j)" for short. Suppose the value of (i, j) for k is of the loglinear form

$$v(i, j, k) = \exp(\mu + L_1(i, j) + L_2(j, k)) \tag{3.1}$$

where μ is an intercept term; $L_1(i, j)$ measures a baseline value of (i, j); and $L_2(j, k)$ shows how the background characteristics of k influence his/her evaluation of j relative to the baseline. This is the simplest model that allows us to handle the application we have in mind, in which the primary interest is to study the factors relating to how different treatments are viewed. The locations are of secondary interest only. More complex models including interactions between i and k, or three-way interactions, could be entertained.

3.1 Effect of Pictures' Characteristics

To make the meaning of (3.1) more concrete, let us assume first that $L_2(j, k) \equiv 0$, so that the background characteristics of a judge would not matter. The simplest model for the effect of treatments j would be a one-way analysis of variance model with $L_1(i, j)=\beta_j$. However, since the location typically has an impact on aesthetic evaluation, this is usually too crude. Two models of practical interest are the *two-way analysis of variance* model

$$L_1(i, j) = \alpha_i + \beta_j \tag{3.2}$$

with $\alpha_I = \beta_J = 0$ (or the picture (I, J) has value μ), and the model with all location-treatment *interactions*

$$L_1(i, j) = \alpha_{ij} \tag{3.3}$$

with $\alpha_{IJ} = 0$. The model (3.2) assumes that the locations and treatments influence preferences multiplicatively. It is also a simple model in which the characteristics of a picture, or location and treatment, are used to explain its attractiveness. The model (3.3) allows for location-treatment interaction. To estimate the model, it is necessary that all pairs (i, j) are involved in at least one pairwise comparison. Under (3.3) the relative value of treatment j, in location i, is given by

$$\frac{\exp(\alpha_{ij})}{\sum_{t=1}^{J} \exp(\alpha_{it})} \tag{3.4}$$

In contrast, under (3.2) the relative value of j would be the same for all i.

For the interpretation of the results under (3.3), it may be the easiest to revert to the commonly used parameterisation

$$\alpha_{ij} = u + u_{1(i)} + u_{2(j)} + u_{12(ij)} \tag{3.5}$$

where $u = \sum_{ij}\alpha_{ij} / IJ$, $u_{1(i)} = \sum_j \alpha_{ij} / J - u$, $u_{2(j)} = \sum_i \alpha_{ij} / I - u$, and $u_{12(ij)} = \alpha_{ij} - u_{1(i)} - u_{2(j)} - u$. Now, $u_{1(i)}$ measures the effect of the location i, $u_{2(j)}$ measures the effect of treatment j, and $u_{12(ij)}$ is the location-treatment interaction.

In general, more complex models could be formulated for the way the characteristics of a picture influence its attractiveness. For example, suppose we have measurements of density D_{ij} of the forest, and average tree height T_{ij} for each picture (i, j). Then we could model the relative value of each picture by taking

$$L_1(i, j) = \beta_1 D_{ij} + \beta_2 T_{ij} \tag{3.6}$$

for example. In Kangas, Karsikko, Laasonen and Pukkala (1993) a similar problem was analysed by first deriving the relative preferences via an eigenvalue calculation, and then using regression to explain the scores in terms of the characteristics of the forest locations. The difference between the two approaches is that (3.6) gives both estimates in a unified setting whereas in the latter approach the estimation of priorities is carried out without any reference to the assumptions needed in regression analysis, and the regression analysis does not take into account that the estimated priorities may differ from the true priorities (cf., Alho and Kangas 1997, 522).

3.2 Effect of Judges' Background Characteristics

Consider $L_2(j, k)$ now. Assume first that the judges can be partitioned into classes $h = 1, \ldots, H$. They can be categories defined by education, social status, place of residence, etc. The goal is to characterise how the classification is related to the way judge k views the treatments j. Define $I_h(k)=1$, if k belongs to class h, and $I_h(k)=0$ otherwise. The simplest model would then be

$$L_2(j,\ k) = \sum_{h=1}^{H} \gamma_{jh} I_h(k) \tag{3.7}$$

where $\gamma_{jH} \equiv 0$ for identifiability. For example, under (3.3) the relative value of treatment j, in location i, is of the form

$$\frac{\exp(\alpha_{ij} + \gamma_{jh})}{\sum_{t=1}^{J} \exp(\alpha_{it} + \gamma_{th})} \tag{3.8}$$

for k.

In analogy with (3.5) we may want to rescale the γ-coefficients so they have mean zero, or we would define $u_{jh} = \gamma_{jh} - \bar{\gamma}_{\cdot h}$, where $\bar{\gamma}_{\cdot h} = \Sigma_j \gamma_{jh} / J$ for $h = 1, \ldots, H-1$. Now the preference of each class h for treatment j can be given in terms of $u_{2(j)}+u_{jh}$.

Instead of categorical explanatory variables, we might consider a continuous explanatory variable Z that would influence a judge's preferences via

$$L_2(j,\ k) = \pi_j Z_k \tag{3.9}$$

This leads to a simple *analysis of covariance* model, and the relative value of (i, j) for k would be

$$\frac{\exp(\alpha_{ij} + \pi_j Z_k)}{\sum_{t=1}^{J} \exp(\alpha_{it} + \pi_t Z_k)} \tag{3.10}$$

In this case (3.4) determines the preferences of those who have $Z = 0$, and π_j shows how the preference for treatment j changes as a function of Z_k. Again, to ease the interpretation, the coefficients π_j can be centred so they sum to zero. Centring the explanatory variables Z may also be useful, because it gives the α_{ij}'s a ready interpretation.

In the practical application of the regression model, we would typically entertain both categorical and continuous explanatory variables. In the simplest case there would be one of each, so for k belonging to $h = 1, \ldots, H-1$ and $j = 1, \ldots, J-1$ we would have

$$v(i,\ j,\ k) = \exp(\mu)\exp(\alpha_{ij})\exp(\gamma_{jh})\exp(\pi_j Z_k) \tag{3.11}$$

under (3.3), for example. Adding other categorical or continuous explanatory variables simply adds new product terms to the formula. It is important to keep in mind that the introduction of Z into the model changes the interpretation given to α_{ij} (and hence to $u_{2(j)}$ and u_{jh}) and γ_{jh}, for example, despite the fact that it enters into (3.11) multiplicatively. The value of (3.10) is a function of Z_k, so the effect of categorical variables on relative priorities varies with the level of Z_k.

It is clear that more complex models can be entertained. For example, we may add interactions between classes h and continuous variables Z. Similarly, we may include interactions between location i and the explanatory variables.

4. DESIGN MATRIX FOR REGRESSION

The practical application of the models of Section 3 is via pairwise comparisons. Let $r\,(i, j, i', j', k)$ be the relative value of (i, j) compared to (i', j') as perceived by judge k. The value is taken to be an estimate of

$$\frac{v(i,\ j,\ k)}{v(i',\ j',\ k)} = \exp(L_1(i,\ j) - L_1(i',\ j') + L_2(j,\ k) - L_2(j',\ k)) \tag{4.1}$$

We see from (3.10) that adding an arbitrary constant to all parameters γ_{jh}, or π_j, will leave the relative value unchanged. Such additive constants would cancel in (4.1). Therefore, some restrictions are necessary to guarantee the identifiability of the parameters from the pairwise comparisons data. We will assume that $\gamma_{Jh}=0$ for all h, and $\pi_J=0$. This means that the parameters γ_{jh} or π_j become *contrasts*: they measure the relative preference of $j = 1, \ldots, J\text{-}1$ as compared to J.

Defining $y(i, j, i', j', k) = \log\ (r\ (i, j, i', j', k))$ we can formulate the regression model

$$y(i,\ j,\ i',\ j',\ k) = L_1(i,\ j) - L_1(i',\ j') + L_2(j,\ k) \\ -L_2(j',\ k) + \varepsilon(i,\ j,\ i',\ j',\ k) \tag{4.2}$$

where the error term has E[$\varepsilon\ (i, j, i', j', k)$]=0 Recall that using the notation of Section 2.2 the equation (4.2) represents one row of the equations $\mathbf{Y}=\mathbf{X}\boldsymbol{\alpha}+\boldsymbol{\varepsilon}$, where the vector $\boldsymbol{\alpha}$ contains all nonzero parameters. As before, the error term can represent stochastically the possible inconsistency of the elicited evaluations. Here the error term may also represent the inevitable simplifications involved in the formulation of the models of type (4.2), either in terms of missing explanatory variables, or in terms of how they are functionally represented.

Each judge k contributes as many rows to the design matrix as the number of pairwise comparisons of (i, j) to (i', j') he/she has made. In general, this number may vary between judges. The rows of the design matrix can now be formulated in three steps. First, the part deriving from (3.2) or (3.3) follows standard patterns of analysis of variance. In the case of (3.3) there are $IJ-1$ columns (cf., (2.2)), for example.

Second, there may be several categorical variables of the type (3.7) with typically varying values of H. Each one of them adds $(H-1)(J-1)$ "γ-columns" into the design matrix corresponding to the parameters $\gamma_{11}, \gamma_{21}, \ldots, \gamma_{J-1,1}; \gamma_{12}, \gamma_{22}, \ldots, \gamma_{J-1,2}; \ldots; \gamma_{1,H-1}, \gamma_{2,H-1}, \ldots, \gamma_{J-1,H-1}$. This part of the rows is formed according to the following rule: each of the $(H-1)(J-1)$ elements is zero except

- if $j \neq j'$ *and* $j = 1, \ldots, J-1$, and k belongs to $h=1, \ldots, H-1$, then the element corresponding to γ_{jh} is 1; or
- if $j \neq j'$ *and* $j' = 1, \ldots, J-1$, and k belongs to $h=1, \ldots, H-1$, then the element corresponding to $\gamma_{j'h}$ is -1.

Third, each of the continuous variables Z adds $J-1$ "π-columns" into the design matrix corresponding to the parameters $\pi_1, \ldots, \pi_{J-1}$. The $J-1$ elements are all zero except

- if $j \neq j'$ *and* $j = 1, \ldots, J-1$, then the element corresponding to π_j is Z_k; or
- if $j \neq j'$ *and* $j' = 1, \ldots, J-1$, then the element corresponding to $\pi_{j'}$ is $-Z_k$.

These rules suffice for all models considered explicitly above. For example, under (3.11) we have the following expectations

$$E[y(i, j, i', j')] = \alpha_{ij} - \alpha_{i'j'} + \gamma_{jh} - \gamma_{j'h} + \pi_j Z_k - \pi_{j'} Z_k \tag{4.3}$$

In this case the design matrix has $(IJ-1)+(H-1)(J-1)+(J-1)$ columns. Suppose $I=J=2$, $H=3$, and consider judge k who belongs to category $h=2$, and has the value $Z_k=2.3$. Then there are six columns. Suppose the comparisons are ordered as (1,1), (1,2), (2,1), (2,2). If judge k has made all the six comparisons, then the first three columns are the same as those in (2.2) for him/her. However, if the judge has not made the comparison (2,1) to (2,2) for example, then the last row would be omitted. The comparison between (1,1) and (1,2) produces the first row as (1,-1, 0,0,1,2.3); a comparison between (1,1) and (2,1) produces the second row as (1,0,-1,0,0,0); a comparison between (1,2) and (2,1) produces the fourth row as (0,1,-1,0,-1,-2.3) etc.

We have written MATHEMATICA functions (Wolfram 1996) that create the design matrix and produce regression estimates, and likelihood ratio and t-test statistics. The programs are available from the authors (osmo.kolehmainen@joensuu.fi).

5. DESIGN OF PAIRWISE EXPERIMENTS

A motive behind using pairwise comparisons rather than direct assessments of the overall priorities, is that the consideration of one pair at the time is expected to reduce biases caused by the ordering of the entities in elicitation. However, only a fraction of all possible pairwise comparisons is often sufficient for the estimation of the ratio scale using regression (Carriere and Finster 1992). A question then arises as to how one should choose the subset of comparisons that will be made. The details depend on the application, and in principle, the whole theory of the planning of experiments for linear models (e.g. Pukelsheim 1993) is available. Here we will merely point out four practical issues.

First, under model (3.3) all pairs (i, j) must appear at least once. We can randomise the order of the photographs, and mark them with labels 1, 2, …, m. The smallest experiment that treats all photographs the same way, is to compare 1 to 2, 2 to 3, ..., and $m-1$ to m. The next smallest experiment would add comparisons 1 to 3, 2 to 4, ..., $m-2$ to m, so the number of comparisons would be $(m-1)+(m-2)$. Larger experiments can be similarly defined. Having chosen the size of the experiment it may be advisable to randomise the order in which the pairs are presented to the judges, so the same photographs do not appear too close together.

Second, under model (3.2) we do not have to make all pairwise comparisons, and we may choose to make systematic evaluations of the treatments j within each of the locations i. Treating locations symmetrically and treatments symmetrically, we can plan the experiment so that the same treatment pairs are compared (e.g., 1 to 2, 2 to 3, ..., $J-1$ to J; and possibly 1 to 3, 2 to 4, ..., $J-2$ to J, etc.) within each landscape. Additionally, we may choose to compare locations using fixed treatments etc. Note that the potential gains in efficiency have then been bought by the *assumption* that (3.2) is correct. In any case, it may be advisable to randomise order of the chosen comparisons.

Third, when there are several judges, we may have the opportunity to use different comparisons with different judges. In particular, if the reduced set of comparisons considered under (3.2) is in use, it may be advantageous to randomise (or deliberately choose) the order of the photographs for each judge separately, so that a larger number of all possible comparisons would appear in the experiment.

Fourth, in some circumstances it is necessary to limit the number of questions to a bare minimum. In such a case one might ask the respondent to first pick out the best alternative (or one among the top choices if the top rank is tied), and then compare the best with each of the remaining ones in turn.

6. APPLICATION AND INTERPRETATION

We have applied the methods outlined above to a problem in landscape planning. The purpose of the study was to estimate how trees influence the relative aesthetic value of different landscapes, and to find out whether the background characteristics of the judges influence their preferences. Digitised photographs of several locations were modified using a computer to reflect different landscape treatments. Evaluating photographs is hard (as indeed are many of the applications one might want to approach via pairwise comparisons; cf. Alho *et al.* 1996), so rather large residual errors were expected.

The data were as follows. We had five locations (close-up scenery, pine forest, water source, golf course, mansion) and six treatments (current state, clear cutting, thinning, removal of lower growth, natural state, and historic state). Two locations had only five treatments, so the total number of photographs was 28. We aimed at a balanced design in which each photograph would appear twice. Hence, the maximum number of comparisons made by each of the 94 judges was 28+27=55, or considerably less than the number of pairs, or 378. However, some judges did not respond to all comparisons, and the total number of comparisons recorded was n=4,928.

Details of the empirical analyses will appear elsewhere (Tahvanainen *et al.* 2001). Here we will merely note some methodological issues that may be of interest in other applications. First, we tested (3.3) against (3.2) using a (likelihood ratio) F-test. The interactions were clearly significant. This means that the treatments do not have the same effect on aesthetic value at each location. The respondents' sex did not have a significant effect on the perceived value of the treatments. However, age did (p=.0002). The relative value of clearcutting increased with age relative to thinning and natural state. This is as one would expect. One qualitative characteristic that was of interest to the researchers was their background relative to the area being considered. Three classes were formed: those living elsewhere, forestry experts, and local residents. Adding the background variable was significant (p=.02). Those living elsewhere differed clearly from local people in that they gave the natural state a higher relative value than clearcutting. The experts agreed with those living elsewhere in their dislike of clearcutting, but no similar preference for the natural state existed. These findings are of methodological interest because the background group was not, by itself, significant (p=.15).

The overall quality of the resulting regression can be described by the square of the multiple correlation coefficient. In this case, it may be estimated by $R^2=1-SSR/SST$, where SST equals the sum of squared scores $y(i, j, i', j', k)$, and SSR equals the sum of squared residuals from regression. In our case we had R^2=.22. As expected, considerable uncertainty concerning the aesthetic value of the photographs remains. As noted above, in some applications one may try to reduce the uncertainty of the estimates via the Delphi technique.

The degrees of freedom for SST are n, or the total number of comparisons. The degrees of freedom for SSR are $n-r$, where r is the number of columns in the design matrix of Section 4. An adjusted measure for the variance explained would then be $R^2_{adj} = 1-[SSR/(n-r)]/[SST/n]$. In our application we had n = 4,928 and r = 42 (with 28 photographs and 6 treatments we had 27 parameters α_{ij} of model (3.3), 2×5 = 10 parameters γ_{jh} of model (3.7) and 5 parameters π_j of model (3.9)), so the modification does not change the results.

The best way to display the relative preferences of the background groups is to plot each group's preferences separately using the parameterisation (3.5). Displaying the groups' preferences by treatment was found to be misleading because it suggests that the relative preferences would be comparable within treatment. The problem may be understood via the following example. Suppose the photographs have a high, but equal value for group *A*. Suppose the photographs have low, but variable values for group *B*, so that photograph 1 has the highest value. Then, comparing the relative values given by groups *A* and *B* for each photograph separately could easily lead the reader of the research report to mistakenly conclude that "group *B* had a higher preference for photograph 1 than group *A*".

7. DISCUSSION

Our experience with practical applications suggests that one advantage of the regression approach lies in its flexibility regarding experimental design and statistical inference. We have here described the regression model for multiple judge data, and provided a Bayesian formulation for multi-objective decision making. The characteristics of the entities being compared, or the background characteristics of the judges, can be used as explanatory variables in the regression analysis of pairwise comparisons data. Both categorical and continuous scale explanatory variables can be applied. Moreover, the regression approach does not require that all possible pairs should be compared, but the theory of the planning of experiments can be utilised.

A variance components model was proposed to analyse uncertainties in judgements in the multiple judge case. The advantage of the variance components formulation is that different sources of uncertainty in judgements can be quantified. The particular formulation of the variance components model can be extended by relaxing the assumptions concerning the random effects. However, this leads to new estimating equations to be solved. In a case of a single judge, the Bayesian analysis of interval judgements was used to measure uncertainty directly. The interval judgements might be useful in multiple judge case as well.

The regression approach provides many other opportunities for refining the analysis of the pairwise experiments. For example, in applications with a large number of judges, it frequently happens that some judges either differ from the average in a radical way, or they do not bother to concentrate on the task and respond at random. The first types of individuals can be very informative. They can be automatically found by emulating techniques that have been developed for the screening of regression data for influential observations. The second type of individuals can be similarly found by computing individual level estimates of the residual error. Both diagnostic checks have been implemented in our regression package.

Empirical findings show that there is often a great deal of uncertainty in the elicited priorities. This is not surprising due to the nature of the assessment tasks. For example, expert judgements can be used to predict relative performance of alternative forest plans for the period of 20 years. Also the prioritisation of decision elements can be difficult in practice. Therefore, it is important to measure and illustrate the uncertainties of the pairwise comparisons data to decision-makers. The regression approach provides a methodology for the task.

ACKNOWLEDGMENT

This research was partially supported by the Finnish Academy of Sciences.

REFERENCES

ALHO, J.M. AND KANGAS, J. 1997. Analysing uncertainties in experts' opinions of forest plan performance. Forest Science 43: 521-528.

ALHO, J.M., KANGAS, J. AND KOLEHMAINEN, O. 1996. Uncertainty in the expert predictions of the ecological consequences of forest plans. Applied Statistics 45: 1-14.

BASAK, I. 1989. Estimation of the multi-criteria worths of the alternatives in a hierarchical structure of comparisons. Communications in Statistics, Theory and Methods 18: 3719-3738.

BASAK, I. 1990. Testing for the rank ordering of the priorities of the alternatives in Saaty's ratio-scale method. European Journal of Operations Research 48: 148-152.

BASAK, I. 1991. Inference in pairwise comparison experiments based on ratio scales. Journal of Mathematical Psychology 35: 80-91.

BOX, G.E.P. AND TIAO G.C. 1973. Bayesian Inference in Statistical Analysis. Wiley, New York, NY.

BUDESCU, D.V., ZWICK, R. AND RAPOPORT A. 1986. A comparison of the eigenvalue method and the geometric mean procedure for ratio scaling. Applied Psychological Measurement 10: 69-78.

CARRIERE, J. AND FINSTER, M. 1992. Statistical theory for the ratio model of paired comparisons. Journal of Mathematical Psychology 36: 450-460.

CRAWFORD, G. AND WILLIAMS, C. 1985. A note on the analysis of subjective judgement matrices. Journal of Mathematical Psychology 29: 387-405.

DE JONG, P. 1984. A statistical approach to Saaty's scaling method for priorities. Journal of Mathematical Psychology 28: 467-478.

DITTRICH, R., HATZINGER, R. AND KATZENBEISSER W. 1998. Modelling the effect of subject-specific covariates in paired comparison studies with an application to university rankings. Applied Statistics 47: 511-525.

GENEST, C. AND RIVEST, L.-P. 1994. A statistical look at Saaty's method of estimating pairwise preferences expressed on a ratio scale. Journal of Mathematical Psychology 38: 477-496.

KANGAS, J., ALHO, J., KOLEHMAINEN, O. AND MONONEN, O. 1998. Analysing consistency of experts' judgements - case of assessing forest biodiversity. Forest Science 44: 610-617.

KANGAS, J., KARSIKKO, J., LAASONEN, L. AND PUKKALA, T. 1993. A method for estimating habitat suitability on the basis of expertise. Silva Fennica 27: 259-268.

LESKINEN, P. AND KANGAS, J. 1998. Analysing uncertainties of interval judgement data in multiple criteria evaluation of forest plans. Silva Fennica 32: 363-372.

LINSTONE, H.A. AND TUROFF M. (eds.) 1975. The Delphi Method: Techniques and Applications. Addison-Wesley, Reading, MA.

LOOTSMA, F.A. 1993. Scale sensitivity in the multiplicative AHP and SMART. Journal of Multi-Criteria Decision Analysis 2: 87-110.

MCFADDEN, D. 1974. Conditional logit analysis of qualitative choice behaviour. Pages 105-142 in Zarembka, P. (ed.) Frontiers in Econometrics. Academic Press, New York, NY.

MCFADDEN, D. 1981. Econometric models of probabilistic choice. Pages 198-272 in Manski, C.F. and McFadden, D. (eds.) Structural Analysis of Discrete Data with Econometric Applications. MIT Press, Cambridge MA.

PUKELSHEIM, F. 1993. Optimal Design of Experiments. Wiley, New York, NY.

SAATY, T.L. 1977. A scaling method for priorities in hierarchical structures. Journal of Mathematical Psychology 15: 234-281.

SAATY, T.L. AND VARGAS, F. 1984. Comparison of Eigenvalue, logarithmic least squares and least squares methods in estimating ratios. Mathematical Modelling 5: 309-324.

SALO, A.A. AND HÄMÄLÄINEN, R.P. 1997. On the measurement of preferences in the analytic hierarchy process. Journal of Multi-Criteria Decision Analysis 6: 309-319.

TAHVANAINEN L., TYRVÄINEN L., IHALAINEN M., VUORELA N. AND KOLEHMAINEN O. 2001. Forest management and public perceptions - visual versus verbal information. Landscape and Urban Planning 53: 53-70.

WOLFRAM, S. 1996. Mathematica Book 3rd ed. Addison-Wesley, New York.

ZAHEDI, F. 1986. A simulation study of estimation methods in the analytic hierarchy process. Socio-Economic Planning Sciences 20: 347-354.

ZHANG, S.-S. AND GENEST, C. 1996. Etude d'un test de confirmation des priorités dans le cadre du procédé d'analyse hiérarchique. Revue de Statistique Appliquée 44: 81-103.

Chapter 16

Using GeoChoicePerspectives in Collaborative Spatial Decision Making

Piotr Jankowski
Department of Geography, University of Idaho, Moscow, ID USA

Timothy Nyerges
Department of Geography, University of Washington, Seattle, WA USA

Key words: Spatial decision support, collaboration, group decision making, site selection, stakeholder, GIS

Abstract: This paper evaluates the use of spatial group decision support software during collaborative decision making in small, inter-organisational groups. To study human-computer-human interaction, our experimental design used a conference room setting and 109 volunteers formed into 22 groups of 5 persons, each containing multiple (organisational) stakeholders. Digital maps were integrated with multiple criteria decision models to select habitat restoration sites in the Duwamish Waterway of Seattle, Washington. Experimental findings demonstrated that groups used maps predominantly to visualise evaluation results and much less to structure/design the decision problem. While the use of multiple criteria decision models by groups remained steady throughout different phases of the decision process, the use of maps was much lower during the initial (deliberative-structuring) phase, than during the later (analytical) phase. Group conflict was higher during the analytical phase and much lower during the deliberative-structuring phase. A higher level of conflict during the analytical phase suggests that analytical decision aids aimed at conflict management are likely to help mitigate conflict, often a necessary part of making progress in public decision problems.

1. INTRODUCTION

Conflict often arises due to peoples' differences in values, motives, and/or locational perspectives about what is to be accomplished (Susskind

D.L. Schmoldt et al. (eds.),
The Analytic Hierarchy Process in Natural Resource and Environmental Decision Making, 253–268.

and Cruikshank 1987, Gray 1989, Gregory 1999). In such situations conflict, and therefore negotiation management with shared decision making, is a fundamental concern in coming to consensus about choices to be made (Susskind and Field 1996, Simosi and Allen 1998). Dealing with locational conflict in an open manner is becoming more important as citizen (stakeholder) participation increases in land use, natural resource and environmental decision making (Parenteau 1988, Crowfoot and Wondolleck 1990, Gregory 1999). The primary rationale for enhanced stakeholder participation in public land planning is based on the democratic maxim that those affected by a decision should participate directly in the decision making process (Smith 1982, Parenteau 1988). In has been said that decision making groups are fundamental building blocks and at the same time agents of change within organisations, communities, and society (Poole 1985). To add to that, Zey (1992 p 22) states "... that decisions [in society] are most frequently made by groups within the context of larger social collectives."

The above perspectives indicate a broad-based need for methodology addressing the needs of group decision making in general and collaborative spatial decision making (CSDM) more specifically. In this chapter we emphasise collaborative decision making that includes computerised decision support.

The need for computerised decision support results from the importance of group decision making and problem solving carried out predominantly during meetings, and from common problems associated with meetings. These problems include: overemphasis on social-emotional rather than task activities, failure to adequately define a problem before rushing to judgement, pressure constricting creativity felt by subordinates in the presence of bosses, and the feeling of disconnection/alienation from the meeting (Nunamaker *et al.* 1993). Other problems hampering the effectiveness of meetings are given by Mosvick and Nelson (1987) and include: ime consumption, inconclusive results, disorganisation, lack of focus, individuals dominating discussion, ineffective for making decisions, and rambling, redundant, or digressive discussion. Despite these negative effects, the attractiveness of a group approach to decision making comes from the fact that individual contributions are increased by a synergistic effect resulting from meeting dynamics.

Methodologies and tools encompassing CSDM employ many different methods. They include critiques of GIS as a construction of positivist thinking, constraining alternative views of reality that otherwise might broaden the decision making discourse (Lake 1993, Sheppard 1995), GIS extensions aimed at improving its decision support capabilities (Densham 1991), group support systems technology as well as theoretical and empirical studies of its use (Jessup and Valacich 1993), work on capturing the dynamics of argumentation (Conklin and Begeman 1989), and research on the human dimensions of groupware and computer networking (Oravec 1996). These approaches contain various viewpoints of decision making that can be described generally as a collaborative and decision analytical perspective. A collaborative approach views decision making as an evolutionary process that progresses from unstructured discourse to problem resolution using discussion, argumentation, and voting. An analytical approach uses quantitative models to analyse structured parts of a decision problem leaving the unstructured parts for the decision makers' judgement. We argue that both approaches are needed in a collaborative decision support environment. To effectively support group participation in decision making, collaboration and decision analysis tools must be integrated to address complex, ill-structured decision situations (Bhargava *et al.* 1994, Stern and Fineberg 1996). As a step towards computerised support tools for group participation in decision making we present GeoChoicePerspectives—a collaborative spatial decision support system, and evaluate its use in group decision making experiment involving a realistic habitat restoration problem. GeoChoicePerspectives is based on the ideas and concepts implemented in a research prototype for collaborative spatial decision support, Spatial Group Choice (Jankowski *et al.* 1997).

To set the context for an evaluation of GeoChoicePerspectives we outline requirements for collaborative spatial decision support software. Following software design requirements, we present the architecture of GeoChoicePerspectives. In section four we evaluate GeoChoicePerspectives using the results of an experimental case study involving a habitat redevelopment problem from Seattle, Washington USA. We conclude the chapter with the discussion of prospects for future development of collaborative spatial decision support systems.

2. REQUIREMENTS FOR COLLABORATIVE SPATIAL DECISION SUPPORT SOFTWARE

Any decision situation involving the collaboration of stakeholders, technical specialists, decision makers, etc., can be addressed by identifying and documenting its various aspects, particularly the ones that influence the decision strategy acceptable to all collaborating parties and its constituent decision tasks. This approach to setting up a decision support environment is called *needs assessment* for decision support (Nyerges and Jankowski 1997). The requirements for computerised support of collaborative spatial decision making depend on the need assessment performed for a decision situation at hand. Among the more important aspects of those needs are understanding the type of collaborators (e.g. novices or experts) and meeting venues for collaborative work (e.g. face-to-face meeting, long-distance conference, different place/different time group work). Despite the potential variability in collaborator and meeting venue types there are, however, sufficiently common tasks such that software can be developed to support a range of participants in the context of various meeting venues. Meeting participants are likely to collaborate on design and construction of various geographical alternatives, sharing interactive mapping tools over a local area network (Faber *et al.* 1995). The evaluation of collaboratively designed alternatives can be carried out with multiple criteria evaluation techniques enhanced by voting tools (Malczewski 1996). The evaluation results can be visualised on special-purpose maps capable of geographically representing consensus solutions (Armstrong and Densham 1995).

Based on the knowledge abilities of decision participants (as they range from experts to novices in using spatial decision support tools) and meeting venues (as they range across place and time), the following design requirements are common.

- A spatial decision support system for collaborative work should offer decisional guidance to users in the form of an agenda.
- A system should not be restrictive, allowing the users to select tools and procedures in any order.
- A system should be comprehensive within the realm of spatial decision problems, and thus offer a number of decision space exploration tools and evaluation techniques.
- The user interface should be both process-oriented and data-oriented allowing an equally easy access to task-solving techniques as well as maps and data visualisation tools.
- A system should be capable of supporting facilitated meetings and hence, allow for the information exchange to proceed among group members, and between group members and the facilitator. It should also support space- and time-distributed collaborative work by facilitating information exchange, electronic submission of solution options, and voting.
- A system functionality should include extensive multiple criteria evaluation capabilities, sensitivity analysis, specialised maps to support the enumeration of preferences and comparison of alternative performance, voting, and consensus building tools.

In the following two sections we present: 1) the architecture of GeoChoicePerspectives—a spatial decision support system developed especially for collaborative meetings and 2) the evaluation of

GeoChoicePerspectives on the basis of its use in habitat redevelopment site selection problem.

3. DESIGN OF GEOCHOICEPERSPECTIVES

GeoChoicePerspectives™ (GCP) software, developed by Geo Choice Inc. of Redmond, Washington USA (http://www.geochoice.com), supports group-based decision making in a geographic information system (GIS) context. Decision participants use GCP to explore, evaluate and prioritise preferences on all aspects of a decision-making process involving multiple criteria and options. Options (decision alternatives) can be represented as points, lines or areas with their attributes (criteria). Multiple perspectives on options evaluation can be combined to provide an overall perspective. Single users can use the GCP to collate multiple evaluations of an option ranking. Groups can use GCP to combine multiple perspectives on criteria and options in an iterative process for consensus building.

3.1 Software Architecture

GCP is composed of three components: GeoVisual™, ChoiceExplorer™, and ChoicePerspectives™ (Figure 1). The GeoVisual™ component is used by decision participants for exploring geographic data on maps, and presenting the results of site option rankings for single user and/or group contexts that are generated by ChoiceExplorer™ or ChoicePerspectives™. GeoVisual™ is implemented as an extension of the ArcView® GIS platform. The ChoiceExplorer™ component is used by decision participants to perform criteria selection and weighting plus options evaluation and prioritisation. ChoicePerspectives™ collates rankings from ChoiceExplorer™ that are subsequently displayed as consensus maps in GeoVisual™. GeoVisual™ and ChoiceExplorer™ are dynamically linked to support interactive computation and display.

The GeoChoicePerspectives™ package can be used in a variety of meeting venues:

- in face-to-face meetings – participants meet at the same place and same time,
- in storyboard meetings – participants meet at the same place, but at different times,
- in conference call meetings – participants meet in different places at the same time,

- in distributed meetings—participants meet in different places at different (convenient) times.

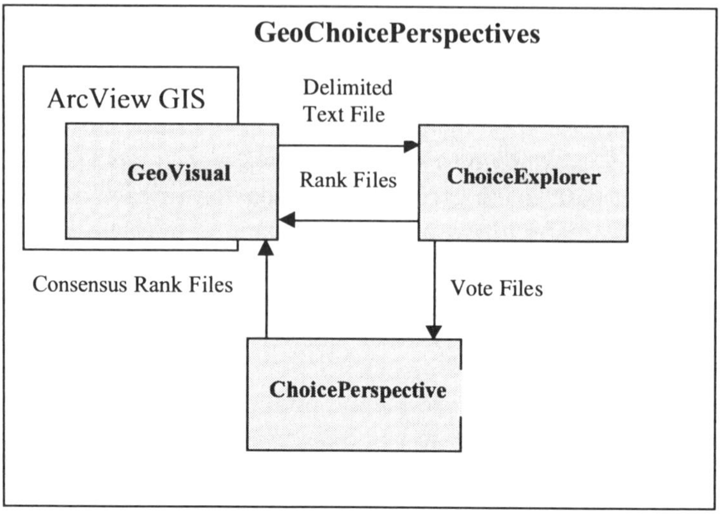

Figure 1. GeoChoicePerspectives software architecture.

Single copies of GeoChoicePerspectives™ can support face-to-face and storyboard meetings (i.e., same place meetings). Multiple copies are needed to support conference call and distributed meetings (i.e., different place meetings).

3.2 Software Capabilities

The functional capabilities of the software include option and option attribute visualisation on a variety of user selected maps, multiple criteria evaluation tools, voting, and consensus building tools.

Option visualisation tools. The nature of the decision options can be reviewed by creating attribute comparison maps (Figure 2). These maps let the user observe and compare numeric information on various option attributes. Background information on decision options can be explored on thematic maps. Option ranks can be presented as a graduated circle map (with site ranking labelled in each circle (Figure 3). Options represented on a map can be hotlinked with documents, photos, and video clips providing additional background information.

Multiple criteria evaluation tools. Properties of evaluation criteria can be set by valuation, standardisation, threshold, and cut-off values (a criterion valuation function lets the software distinguish among benefit, cost, and range criteria). Criterion weights can be assigned using AHP-based pairwise comparison (Figure 4), ranking, and rating techniques. Ranking of decision options can be generated using one of three aggregation functions (decision rules): weighted summation, ordinal ranks, and ideal point. The user can explore the 'robustness' of the ranking to changes in criterion weights by performing sensitivity analysis (Figure 5).

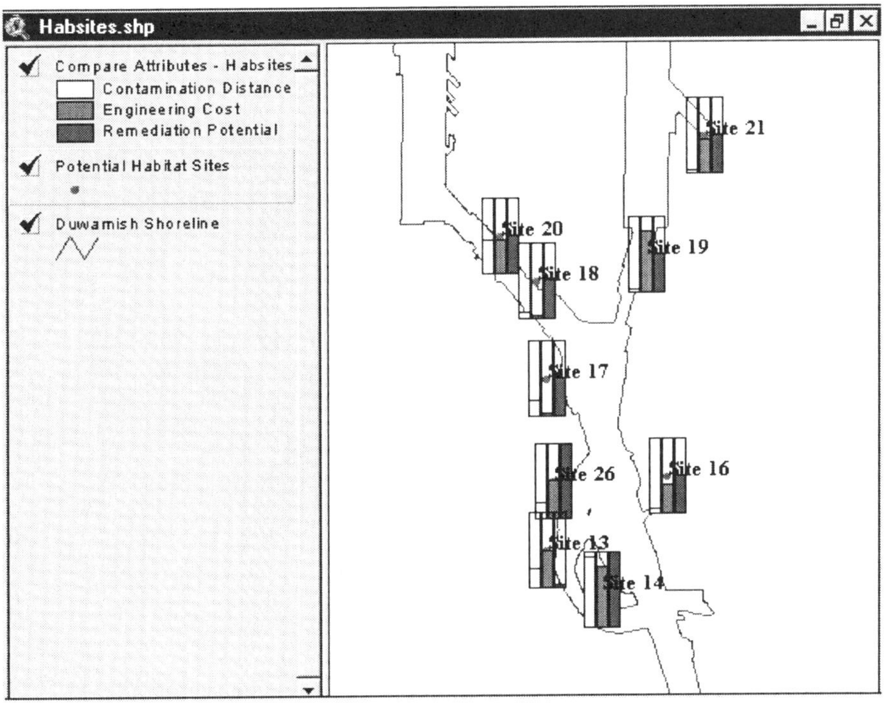

Figure 2. A multiple histobar map allows one to compare attribute values and aids the pairwise comparison weighting method.

Voting tools. The participants can vote electronically on the choice of criteria, cut-off values, threshold values, standardisation method for valuation of criteria, criterion weighting method, criterion weights, aggregation function, and the order of ranked decision options (Figure 6). They can even vote on some more specific and personal aspect of the decision problem by using the generic vote feature.

Consensus building tools. Consensus mapping can be used to communicate which options are the best ones. The map (Figure 7) uses circle size to represent overall group preference including the group ranking (based on Borda score) and the variance of the ranking. Participant votes can be aggregated using non-ranked and ranked methods. A non-ranked method is the standard "simple majority count" (in other words, the voter gives all items (e.g. options) included in the vote an equal preference). A ranked vote method takes into consideration the order of significance of what is being voted upon (such as options).

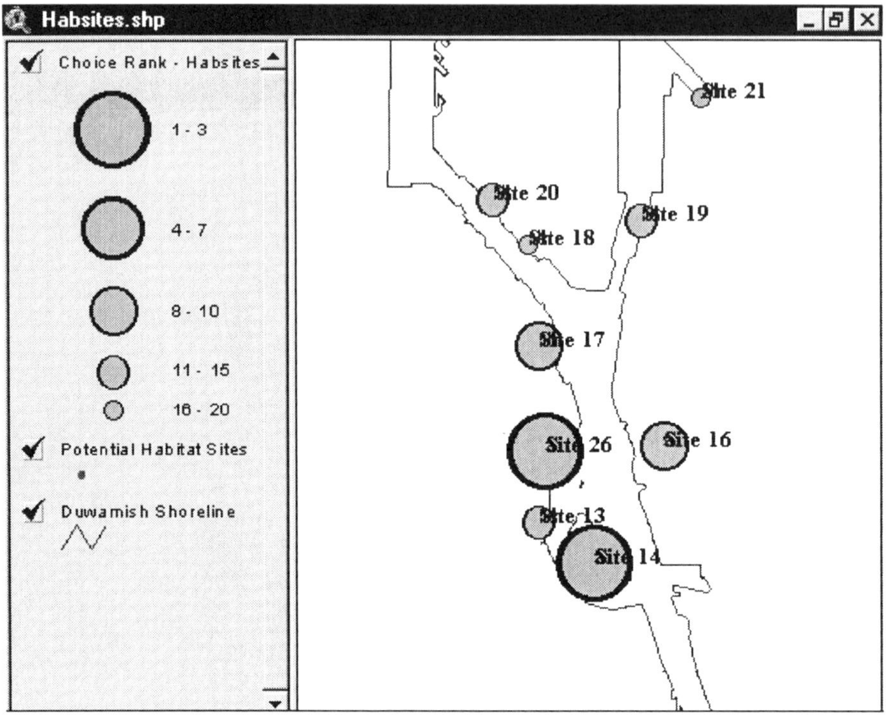

Figure 3. Option rank map uses rank-graduated circle size to display ranking results and option locations to help visualise spatial relationships.

4. EVALUATION OF GEOCHOICEPERSPECTIVES: A HABITAT RESTORATION DECISION PROBLEM

We used GeoChoicePerspectives in the experimental study of habitat site selection along the Duwamish Waterway in Seattle. The study was composed of a laboratory experiment setting in which we were able to videotape participants in groups working with computer-oriented geographic information. The socio-behavioral setting involved groups of 5 participants assisted by a facilitator/chauffeur using GeoChoicePerspectives groupware software in a decision laboratory. Our choice of 5-person groups stems from Vogel's (1993) review of several experiments in GSS research that showed mixed results with groups of 3 or 4, but beneficial results starting with a group size of 5. The study used 109 participants formed into 22 groups (one group had only four members). They were recruited from across the University of Washington campus, and a few from off campus, through announcements in classes and flyers posted on bulletin boards around campus. The average age of the participants was 28 years. The average education attainment was close to completion of an undergraduate degree, although there were several graduate students and participants from off-campus with an interest in GIS and habitat restoration.

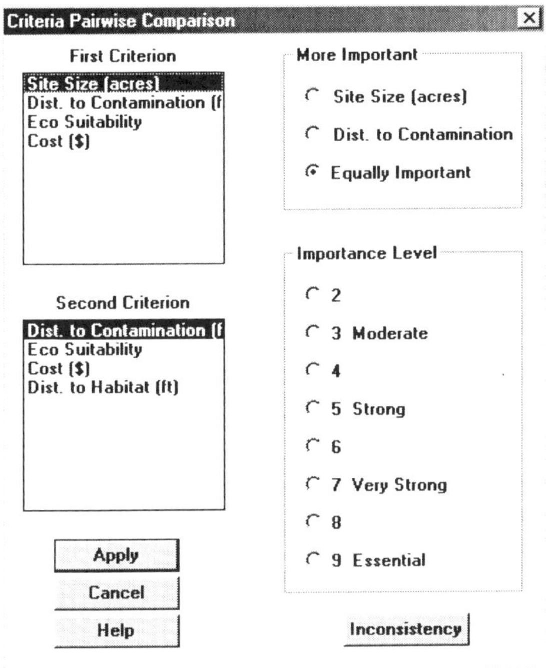

Figure 4. Pairwise comparison criterion weighting based on AHP method (Saaty 1980).

4.1 Decision Task

We adopted a realistic decision task to structure our treatments about site selection for habitat restoration (development) in the Duwamish Waterway of Seattle, Washington. The decision task was being performed by the National Oceanic and Atmospheric Administration (NOAA) Habitat Restoration Panel (NOAA 1993) due to a law suit settled against the City of Seattle and King County for inappropriate storm sewer drain management. For years storm sewer drains had been releasing unfiltered storm water containing highway gasoline and oil contaminants into Puget Sound (Elliott Bay) degrading the fish and wildlife habitat. A GIS database for site

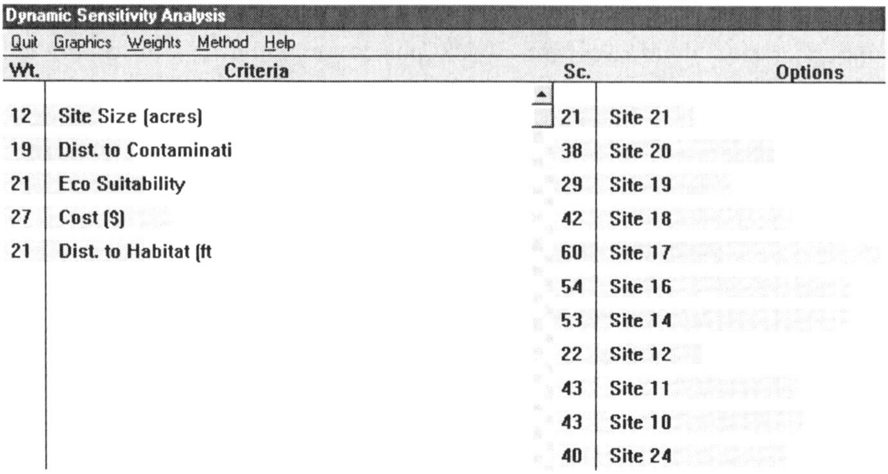

Figure 5. Dynamic sensitivity analysis allows evaluating changes in the ranking in response to changes in criterion weights.

selection problem was compiled from City of Seattle and King County sources and included 20 sites (Figure 8). The site selection decision process was expected to involve conflict management during social interaction due to the different perspectives inherent in the views of participating members. Thus, site selection activities are particularly interesting from the standpoint of software tool use and its interplay with group interaction.

Each decision group met for five meeting sessions, one in each of five consecutive weeks (or as close as possible to that schedule), and worked on a different version of the habitat site-selection task. In each of the five sessions we asked the groups to work toward consensus on the selection of three preferred sites (or as many as the $12 million budget would allow) out of the total number of sites presented to them. The total number of sites varied from eight to twenty. At the end of each session, we asked a group to fill out a session questionnaire which provided a means for the individuals to assess group use of the tools, group interaction, and the level of satisfaction with the overall group selection.

Data were collected by session (hence task) using questionnaires and coding interaction of videotapes. Each participant filled out a background questionnaire (education, sex, age, etc.) and attended a two-hour CSDM software training session. At that time, we passed out materials introducing the overall wildlife habitat site-selection task, assigned the participants to groups based on schedule availability, and handed out stakeholder roles that they could adopt by the time their first decision session convened. Based on interviews completed by the NOAA Restoration Panel (NOAA 1993), participants could adopt a role of business/community leader (20 adopted it), elected official (10 adopted it), regulatory/resource agency staff member (22 adopted it), technical/academic advisor (23 adopted it), or environmental group representative (29 adopted it). Roles were self-selected to encourage subjects to participate based on their inherent interests. We made sure that no less than three different stakeholder roles were represented in each of the groups.

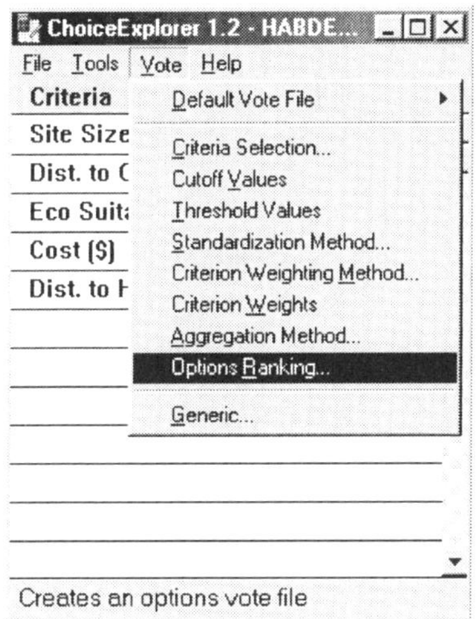

Figure 6. Voting tools enable the participants to vote on every aspect of criteria selection and option ranking process.

4.2 Experiment Results

In regards to our experiment results in particular, we found that background thematic maps were used predominantly to visualise the locations of decision alternatives and could potentially be used to evaluate trade-offs among the decision alternatives. We were surprised to see that special purpose maps (option rank map—Figure 3, and consensus map—Figure 7) designed to facilitate conversations about evaluation and prioritisation were used not as much as we would have thought—in less than 5% of the moves to invoke the use of maps. We speculate that background maps showing site locations are easier to interpret than the special purpose maps that try to combine location and ranking. There is no evidence in this experiment to suggest that maps included in GeoChoicePerspectives software were effective in prioritising evaluation criteria, displaying the results of sensitivity analysis, and the position of a group in regard to the final ranking of decision alternatives. Groups used maps predominantly to visualise the evaluation results and much less to structure/design the decision problem. The high frequency of map moves for situation maps and orthophoto images, especially during the decision phase involving the evaluation/selection of alternatives, shows the usefulness of reference maps (i.e., both a general situation map and a realistic orthophoto image) in presenting the results of decision alternative rankings. The question arises

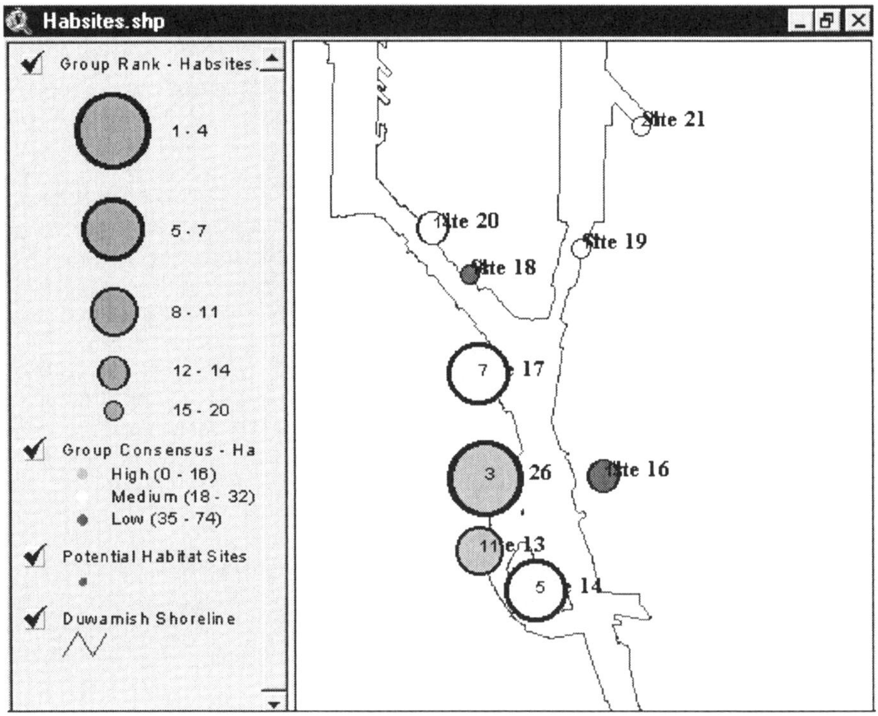

Figure 7. Consensus map uses circle size and colour to display the group ranking vote results. The larger the circles the higher the rank scores. Green colouring indicates relatively higher consensus for those options, yellow indicates relatively medium, and red indicates relatively low consensus.

then—were the maps provided in GeoChoicePerspectives simply not adequate for problem exploration, criteria identification, valuation, and prioritisation? Based on the analysis of variance, maps implemented in GeoChoicePerspectives played only a limited support role in the decision stages of the experiment. How to improve the existing maps and which direction should be taken in the design of new types of maps and visualisation aids are open research questions. In regard to overall map use and facilitation, there was a noticeable difference in the mean frequency of map use between task 4 (a facilitated session that included both individual and public display) and task 5 (public display only supported by a facilitator). Testing to see if the influence of a facilitator can help with the interpretation of this finding is one way to go. Testing a variety of map displays from simple to more complex is another approach.

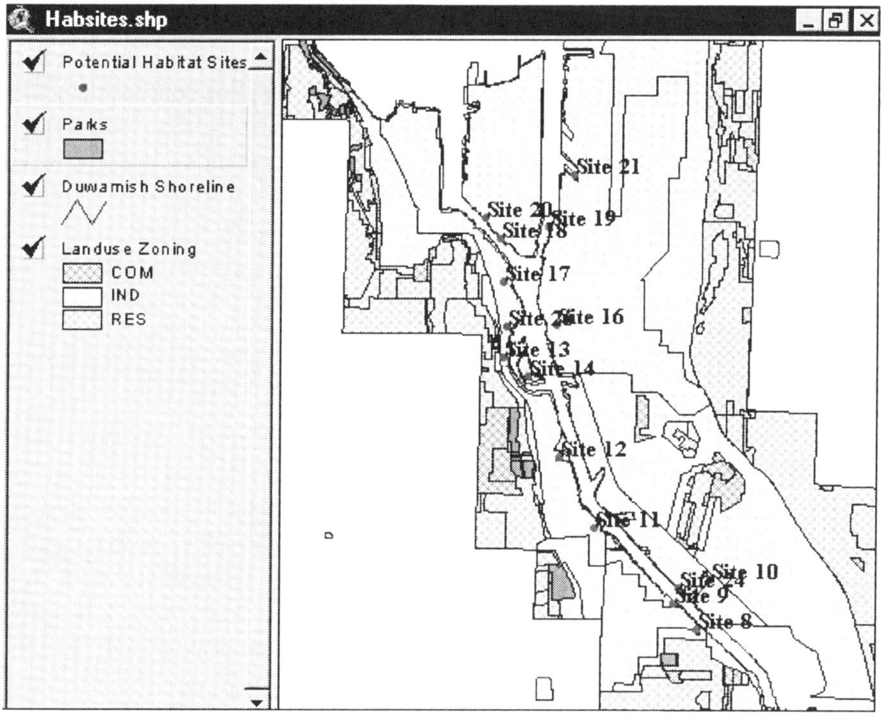

Figure 8. Potential habitat redevelopment sites in the Duwamish Waterway

Decision process-oriented findings were somewhat intriguing. It was surprising to find that the participants used multiple criteria evaluation tools without much difference in the frequency of moves in both halves of the experiment, in which they engaged in multiple criteria-based evaluation. We speculated that the first, more "exploratory" half of the experiment would be marked by more frequent use of maps than the second half. The much less frequent use of maps during the first half of experimental sessions indicates a re-examination of the exploratory usefulness of maps is needed. We also found that different phases of the decision process had two different levels of conflict: analytical detail phase characterised by a high level of conflict and exploratory-structuring phase characterised by a low level of conflict. There was less conflict during problem exploration because interests and values were not at odds with each other. There was more conflict during criteria selection and alternative evaluation because interests showed up here. The higher level of conflict during the evaluation phase tells us that analytical decision aids aimed at conflict management are likely to help move through conflict; such conflict now being recognised as a necessary part of making progress in environmental disputes. Future designs of collaborative spatial decision support software should take this into consideration, and provide capabilities to manage conflict more directly.

When it came to task complexity, our findings were a bit surprising as well. Task complexity was not associated with the level of conflict between tasks 1 (simpler) and task 4 (more complicated), a finding somewhat contrary to current literature. Other factor differences such as task 5 with public-only display versus task 4 with public-private displays showed differences in conflict. However, whether the difference is due to opportunity to voice opinion or due to conflict over what to display is not clear.

5. SUMMARY AND CONCLUSIONS

Many spatial decision problems of a participatory nature are likely to involve conflicting perspectives on facts, interests, as well as world views (Renn *et al.* 1995). Together these differences add to the complexity of trying to come to agreement. Software systems such as GeoChoicePerspectives (GCP) are not expected to "generate" the consensus, but only help in the negotiation of shared understandings that lead to agreements. Many reports in the literature indicate that conflict is a necessity in complex, participatory decision making. Conflict is necessary to sort through the differences in facts, interests, and world views (Renn *et al.* 1995). Only after such conflict arises might there be a chance for integration of the differing aspects, promoting a shared understanding of differences, and perhaps subsequent agreement.

The description of GCP provided here focused on the technical aspect as it related to the habitat decision situation. The habitat decision situation described above consisted of only one major task—that of option evaluation—with a series of subtasks. The criteria were identified, and the basic set of options was provided. This masked the fact that multiple stakeholders were interviewed to gain an understanding of what was of concern. It also masked the fact that a select few decision participants generated the initial set of options for site selection, itself a group process. To highlight a different approach, Renn *et al.* (1995) described an energy policy process that looked the same, consisting of three steps, but was very different because a different "culture-focused" group was used for each phase of a participatory decision process: 1) values and criteria elicitation was undertaken by stakeholder groups, 2) option generation was performed by a group of technical specialists, and 3) options evaluation was performed by a randomly selected group of the general public. Each of those phases is likely to have a different dynamic about it, and hence system requirement to help sort through the nature of disagreements. Providing technology to support each of the different phases, taking into consideration the different groups that might be involved, and documenting the results of each phase is a very important part of the transparency over the process. Developing information technology that takes into consideration easy access to analytic results, and highlighting the commonality and differences in perspectives, requires further integration of collaboration technologies and GIS technologies. Such integration is the likely development direction of collaborative spatial decision support systems.

REFERENCES

ARMSTRONG, M.P. AND DENSHAM, P. J. 1995. Cartographic support for collaborative spatial decision making. Pages 48-58 in Proceedings of Auto-Carto 12. American Congress on Surveying and Mapping, Bethesda MD.

BHARGAVA, H. K., KRISHNAN, R. AND WHINSTON, A. D. 1994. On integrating collaboration and decision analysis techniques. Journal of Organizational Computing 4:297-316.

CONKLIN, J. AND BEGEMAN, M. L. 1989. gIBIS: A tool for all reasons. Journal for the American Society of Information Science 40:200-213.

CROWFOOT, J. E. AND WONDOLLECK, J. M. 1990. Environmental Disputes: Community Involvement in Conflict Resolution. Island Press, Washington, D. C.

DENSHAM, P. J. 1991. Spatial decision support systems. Pages 403-412 in Maguire, D. J, Goodchild, M. F, and Rhind, D. W. (eds.) Geographical Information Systems: Principles and Applications. John Wiley & Sons, New York.

ELECTRICAL POWER RESEARCH INSTITUTE. 1998. Available at: http://www.smartplaces.com.

FABER, B., WALLACE, W. AND CUTHBERTSON, J. 1995. Advances in collaborative GIS for land resource negotiation. Pages 183-189 in Proceedings of the GIS'95 Ninth Annual Symposium on Geographic Information Systems. GIS World, Vancouver, B. C.

GRAY, P., VOGEL, D. AND R. BEAUCLAIR. 1990. Assessing GDSS empirical research. European Journal of Operational Research 46:162-176.

GREGORY, R. 1999. Identifying environmental values. Pages 32-58 in Dale, V.H., and English, M.R. (eds.) Tools to Aid Environmental Decision Making. Springer-Verlag: New York.

JESSUP, L. AND VALACICH, J. (eds.) 1993. Group Support Systems: New Perspectives. Macmillan Publishing Company, New York.

JANKOWSKI, P., NYERGES, T. L., SMITH, A., MOORE, T. J. AND HORVATH, E. 1997. Spatial Group Choice: a SDSS tool for collaborative spatial decision-making. International Journal of Geographic Information Systems 11:577-602.

LAKE, R. W. (ed.) 1987. Resolving Locational Conflict. Rutgers Center for Urban Policy Research, New Brunswick NJ.

MALCZEWSKI, J. 1996. A GIS-based approach to multiple criteria group decision making. International Journal of Geographical Information Systems 10:955-971.

MOSVICK, R. AND NELSON, R. 1987. We've Got To Start Meeting Like This: A Guide To Successful Business Meeting Management. Scott, Foreman, Glenview Ill.

NOAA (National Oceanic and Atmospheric Administration). 1993. Technical notes from the Elliot Bay/Duwamish restoration program. NOAA Restoration Program, Sand Point Office, Seattle WA.

NUNAMAKER, J. D. A., VALACICH, J., VOGEL, D. AND GEORGE, J. 1993. Group support systems research: Experience from the lab and the field. In Jessup, L. and Valacich, J. (eds.) Group Support Systems: New Perspectives. Macmillan Publishing Company, New York.

NYERGES, T AND JANKOWSKI, P. 1997. Enhanced adaptive structuration theory: a theory of GIS-supported collaborative decision making. Geographical Systems 4:225-259.

ORAVEC, J. 1996. Virtual Individuals, Virtual Groups. Press Syndicate, Cambridge, UK.

PARENTEAU, R. 1988. Public Participation in Environmental Decision-making. Federal Environmental Assessment Review Office, Ottawa.

POOLE, M. S. 1985. Tasks and interaction sequences: a theory of coherence in group decision-making interaction. Pages 206-224 in Street, R. L. Jr., and Cappella, J.N. (eds.) Sequence and Pattern in Communicative Behaviour. Edward Arnold, London.

RENN, O., WEBLER, T. AND WIEDEMANN, P. 1995. Fairness and Competence in Citizen Participation: Evaluating Models for Environmental Discourse. Kluwer Academic Publishers, Dordrecht.

SAATY, A.L. 1980. The Analytic Hierarchy Process. McGraw-Hill, New York

SAGE, A. 1 991. An overview of group and organizational decision support systems. IEEE Control Systems Magazine August:29-33.

SHEPPARD, E. 1995. GIS and society: towards a research agenda. Cartography and Geographic Information Systems 22:5-16.

SIMOSI, M. AND ALLEN, P. T. 1998. Public perception of risk management in environmental controversies: a UK case study. Risk: Health, Safety and Environment 9:309-327.

SMITH, L. G. 1982. Alternative mechanisms for public participation in environmental policy-making. Environments 14:21-34.

STEINITZ, C. 1990. A framework for theory applicable to the education of landscape architects (and other design professionals). Landscape Journal 9:136-143.

STERN, P. C. AND FINEBERG, H. V. (eds.) 1996. Understanding Risk: Informing Decisions in a Democratic Society. National Academy Press, Washington, D. C.

SUSSKIND, L. AND CRUIKSHANK, J. 1987. Breaking the Impasse: Consensual Approaches to Resolving Public Disputes. Basic Books, New York.

SUSSKIND, L. AND FIELD, P. 1996. Dealing with an Angry Public. The Free Press, New York.

ZEY, M. (ed.) 1992. Decision Making: Alternatives to Rational Choice Models. SAGE, Newbury Park CA.

Chapter 17

Integrating the AHP with Geographic Information Systems for Assessing Resource Conditions in Rural Catchments in Australia

Robert M. Itami and Glen MacLaren
Centre for GIS and Modelling, Department of Geomatics, University of Melbourne, Victoria, Australia

Kathleen Hirst
GIS Applications Pty Ltd., South Caulfield, Victoria, Australia

Key words: Priority setting, multi-criteria analysis, geographic information systems, land capability analysis

Abstract: Decision making in catchments is inherently complex and spatial in nature. This chapter examines the nature of this complexity, proposes criteria for good decision making, and focuses on the utility of the analytical hierarchy process (AHP) as a decision making tool in this context. The AHP has great potential value in integrating qualitative judgements with scientific information. However, its limitations for ranking issues that have a spatial dimension can be resolved by linking the AHP process to geographic information systems (GISs). A computer program called Catchment Decision Assistant provides a friendly user interface to construct AHP hierarchies and generate weighted map overlays using ArcView GIS with the Spatial Analyst extension. The program automates the GIS processing so the user needs to know little about GISoperation . This chapter describes the Catchment Decision Assistant architecture, how the AHP is implemented in a spatial context, and demonstrates its use to assess biophysical capability for horticultural crops in West Gippsland Catchment Management Authority in Victoria, Australia.

D.L. Schmoldt et al. (eds.),
The Analytic Hierarchy Process in Natural Resource and Environmental Decision Making, 269–287.

1. INTRODUCTION

The concept of integrated catchment management presumes a holistic approach to decision making. As natural drainage ways, catchments contain a complex interaction between natural processes and human activities. The idea of sustainability is imbedded in the concept of integrated catchment management. The overall goal of integrated catchment management is to sustain a healthy ecosystem for the enrichment, health and well being of future generations. This is not a trivial task, commanding the best scientific knowledge, the experience gained from local knowledge, and the creativity and vision of the community as a whole.

The task would be difficult enough if we were starting from a healthy ecosystem. Unfortunately, because of a poor understanding of Australian natural processes, past generations have unwittingly made decisions that have led to the gradual degradation of the land as is evidenced by problems of erosion, soil salinity, acidification, and declining water quality. This has led to declining productivity, more costly inputs to farm practices and the inevitable erosion of farm incomes, personal stress and the decline of rural economies.

Where does one start? The scientific community is quick to assist but are the first to admit that the complexity of ecosystems means that no comprehensive understanding of all the cause and effect relationships between natural and social processes can be modelled with certainty. Even if this understanding existed, the lack of data available on natural and social conditions across any given catchment in Australia would render these models useless. Politicians will respond to the will of their constituents, but often the community can easily recognise problems but have great difficulty in identifying a clear set of solutions that adequately address the complex inter-relationships of ecosystems.

What is needed is a comprehensive, integrated approach to decision making that helps people structure these complex problems in a framework that takes advantage of the best scientific knowledge where this exists, and capitalises on the knowledge and experience of local experts and the community. Fortunately, complex decisions are not isolated to catchment management. It is a problem for business, governments, planning, design, and marketing. In fact, complex decision making is a problem that confronts humans on a day to day basis. Purchasing a house or car, playing a game of chess, or driving through busy traffic are all common activities that require complex decision making. The difference in catchments is that management decisions are not made by one person, but by many people from a wide range of backgrounds.

Human decision making involves taking into consideration a range of subjective and objective issues. In reality, human decision making is rarely objective, and usually inconsistent. Yet, society demands accountability and transparency. If public trust is to be maintained, managers and policy makers must take the business of decision making seriously.

One approach that has the capacity to integrate both objective and subjective criteria in the decision making process in a way that is easy for lay people to understand is the analytic hierarchy process or AHP. The analytic hierarchy process is a technique developed by Saaty as a generalised method for dealing with "fuzzy" issues in decision making. Saaty (1995) developed the method in the 1970s. The AHP has been applied in a broad range of environmental impact assessments, catchment management planning, land use planning, and natural resource studies (Banai-Kashani 1989, 1990, Jankowski and Richard 1994, Xiang and Whitley 1994, and Bantayan and Bishop 1998). It has also been employed for over a decade in business and government (see bibliography by Golden *et al.* 1989) to assist in setting priorities and ranking preferences among alternative actions. The AHP presumes the use of computers to handle the mathematical complexities while providing a relatively simple method for user's to express preferences for complex issues.

The AHP has been implemented as one of the decision support tools in the Department of Natural Resources and Environment's Catchment Management Decision Support System (Itami *et al.* 1999). A software package called Catchment Decision Assistant (CDA) implements all aspects of the AHP and records results in a database so weightings and criteria may be reviewed, revised, and reused. In addition the AHP has been adapted so it can be used to select sites by weighting map criteria. In this context the AHP can be used to weight environmental factors that may contribute to a crop such as wine grapes. When these weights are applied to maps in a geographic information system, a map ranking all sites for wine grape production is generated.

2. CATCHMENT DECISION ASSISTANT

The Catchment Decision Assistant software is written in Visual Basic and acts as a user interface to ArcView and Spatial Analyst by ESRI. CDA is one component of a large decision support system for catchment management developed in a cooperative agreement between the Department of Natural Resources and Environment, Victoria and The Centre for GIS and Modelling in the Department of Geomatics at the University of Melbourne (Itami *et al.* 1999). CDA implements all aspects of the AHP and records results in an Access database so weightings and criteria may be reviewed, modified (if required), and reused.

CDA provides a systematic framework for weighting criteria contributing to catchment issues. CDA is available in two versions (see Figure 1). The first version is aimed primarily at decision making at the state-wide level to rank project proposals for funding and does not link to Geographic Information Systems (GIS). The second version is fully integrated with ESRI's ArcView GIS and is designed to simplify access to the large GIS data holdings of the Department of Natural Resources and Environment (DNRE) in the State of Victoria, Australia, and to systematically apply the AHP to prioritise sites using multiple criteria. The first version of the software uses a similar interface to the map version, but instead of a GIS model, produces a Microsoft Excel spreadsheet that can then be used in a committee environment to rank projects using weights and criteria generated by the AHP process. Since the project version of CDA works in a similar fashion to existing software that implements the AHP, the remaining discussion will focus on the GIS version of CDA.

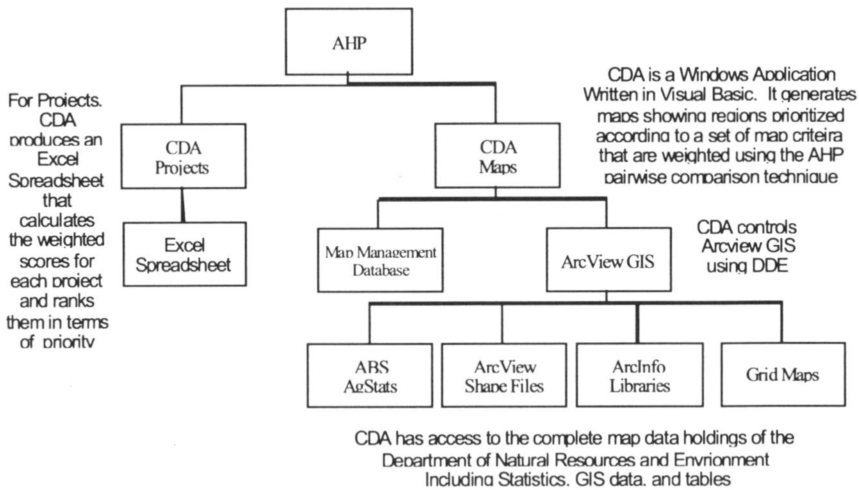

Figure 1. Two versions of the Catchment Decision Assistant (CDA) software have been developed. One that works with ArcView GIS to integrate the AHP with map databases, and the other version that works for non-spatial decision making.

3. CDA SOFTWARE REQUIREMENTS

Target audiences for the CDA software are resource analysts, catchment managers, and community groups. Most people in these groups are unfamiliar with formal decision making processes or concepts of GIS. In addition, most of these users do not have the time or patience to learn the theoretical or technical issues relating to decision making processes or GIS. Therefore any software that is to have any utility in this environment should clearly separate the decision making process from the complexities of GIS operation. In fact, it is desirable to have the GIS "invisible" to the user if possible. Desktop GIS systems such as MapInfo or ArcView go a long way to simplifying the operation of a GIS, however if one is to implement a high level analytical technique such as the AHP, then, the task of operating the GIS can be onerous even to the highly trained expert. CDA resolves these problems by developing a three-tier architecture.

4. CDA THREE-TIER ARCHITECTURE

4.1 Bottom Tier - Database Layer

At the bottom tier are the GIS databases. The GIS databases are complex in themselves, containing over 350 map layers covering all or parts of Victoria at different scales. They are comprised of Population and Housing statistics, agricultural statistics, databases on soils, water, climate, vegetation, wildlife and topography, utility infrastructure, political boundaries. These layers are at different levels of detail, and in different formats including ArcInfo libraries, ArcView shape files, grid libraries, as well as regional datasets with unique geographic boundaries. Naming conventions for these files are often cryptic and impossible to interpret by lay people. To simplify access to the GIS databases, two lookup tables are provided. The first look up table categorises maps into common themes or subject areas. Once the user selects a subject area, individual maps within that theme are displayed. Finally the fields for each map coverage are displayed with a look up table describing each field.

To simplify access to the different data formats, a set of GIS and database access tools have been developed to provide a uniform set of methods for accessing any dataset. These tools are built from a combination of Visual Basic programs to access system database files and lookup tables, and ArcView avenue scripts to manage GIS database files.

4.2 Middle Tier - AHP Modelling

The middle tier of CDA is a set of object oriented tools for querying the GIS database, reclassifying map fields, calculating AHP weights and consistency ratios, and finally generating the final AHP map by generating an ArcView Avenue script to execute the hierarchy. In addition, a set of reporting tools are available to the user to automatically generate a document that describes the criteria, an HTML version of the report with hyperlinks to the associated maps.

4.3 Top Tier - User Interface

The top tier of CDA is the user interface. The interface is written in Visual Basic and is written to hide the complexities of GIS operation from the user. In fact, the user is generally unaware that ArcView is running in the background. CDA uses Dynamic Data Exchange (DDE) to communicate to ArcView. ArcView is used to display maps, generate tabular summaries, and to execute the final model. The next section describes the user interface in more detail. The main components of the user interface include:

- Defining the region of interest
- Building the Decision Hierarchy
- Assigning properties to each criterion in the decision tree.
- Generating weights for criteria using pairwise comparisons.
- Generating the results

4.3.1 Defining the region of interest

The first problem that needs to be addressed when integrating the AHP with GIS is to define the region of interest. Victoria is divided into nine Catchment Management Authorities (CMA) and one Catchment Land Protection Board (CLPB) (Figure 2). Since CDA is designed to address the needs of the CMA's, the database has been organised so the user can select any CMA or CLPB as the region of interest. Once this selection is made, all GIS operations are applied to this region.

4.3.2 Building the decision hierarchy

CDA provides a graphic interface for building the AHP decision hierarchy. Figure 3 shows the form, which includes a toolbar for inserting and deleting criteria. The entire hierarchy along with its weights and intensity ratings for each criterion are stored in a database. The hierarchy may then be reloaded, reviewed, and edited. Hierarchies may also be copied and modified. This is important if, in a group decision making setting, two groups diverge in either the criteria or weights for criteria. In this case the hierarchy can be copied and the two groups can develop their own weights and the results of the two groups subsequently compared.

Figure 2. Victoria's nine Catchment Management Authorities and one Catchment and Land Protection Board. (These are the administrative boundaries for Catchment management in the State of Victoria.)

4.3.3 Assigning properties to each criterion in the decision tree

As the decision hierarchy is built, the users can further define the definitions of the criteria and the AHP Classes. A tabbed form that appears when the user double clicks on criteria on the decision tree in Figure 3. The form allows for descriptions, and notes to be taken. In this way important definitions or points of discussion can be made during the process of defining criteria. These points can then be followed up on a later date or remain as a permanent part of the documentation of the decision process.

If the criteria are mapped, the user then selects the map theme and data fields using the form in Figure 4. This interface has gone through many revisions to make this process as easy and intuitive as possible. Since there are over 300 map themes to select from, the objective is to simplify selection as much as possible. To facilitate this a database containing all map themes has been created to store the map scale, a description, a list of associated look up tables, and the location of the file (map databases may reside on different servers). The user first selects one of 16 subject areas. Next, the user may select a map from this category. Once the map is selected, the user selects an attribute field from the associated table. Fields are generated by a query to ArcView. Once the field is selected, the user may browse through the field values with descriptions if a look up table is available. To confirm the selection, the user may then view the map with legend on the screen.

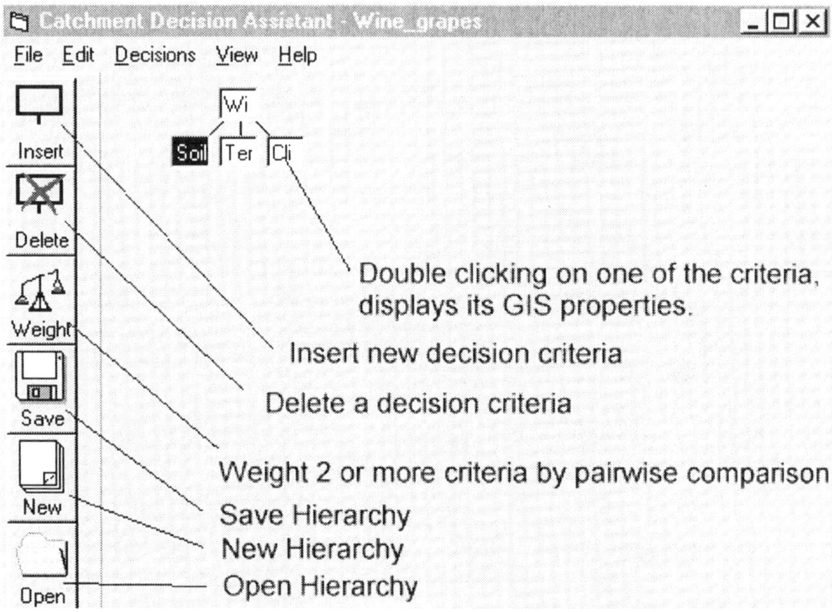

Figure 3. User interface for building the AHP decision hierarchy. The toolbar to the left allows the user to easily insert, delete, or weight decision criteria. By double clicking on any criteria, the user can define the properties for each criterion.

Once a map field is selected, the next step is for the user to define AHP classes. This step defines the intensity ratings for each category in the map field. There are two types of classification. For discrete or categorical data such as soil types, the user may assign ratings to each soil type as is shown in the example in Figure 5. For continuous data such as slope, elevation, or temperature, the user can either define discrete classes by dividing the range into equal intervals or by user defined classes. No matter which method is selected, the user then assigns intensity ratings in the range from zero to one. These AHP classes provide the definitions for ratings that are then applied to the field values.

The final step in defining map properties for a criterion is to assign the AHP Class ratings to the map fields.

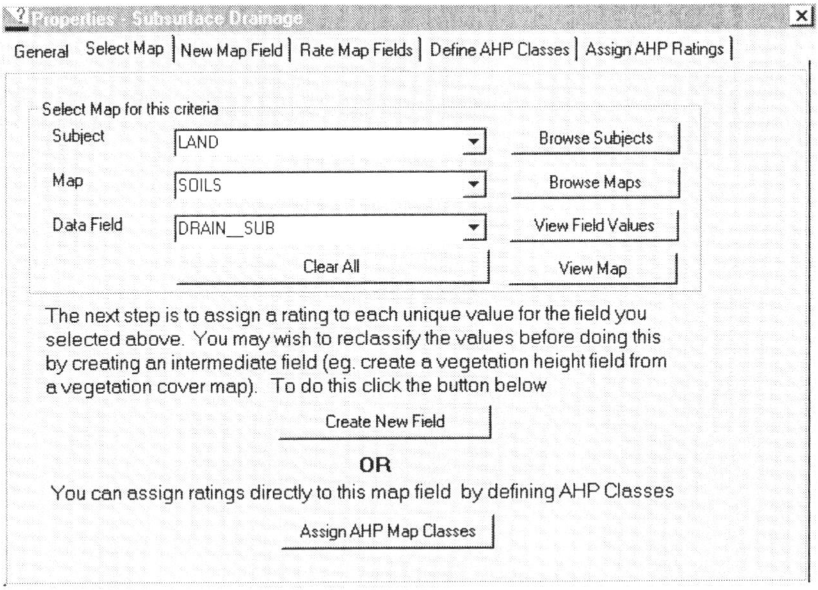

Figure-4. If the criterion is mapped, the user selects the map (grid, shape file, or coverage) using the "Select Map" tab on the properties form. The user selects the map by subject, then the lists of maps for that subject is displayed, and finally once a map is selected the associated fields in the associated table are displayed.

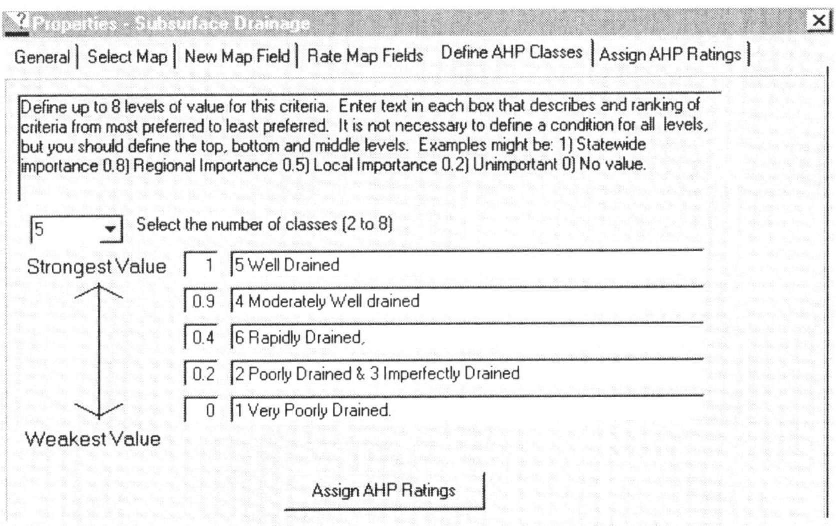

Figure 5. Once the map field is selected, continuous values may be classified using equal interval or user selected class boundaries. Resulting classes are then assigned intensity ratings from 0 to 1.

4.3.4 Generating weights for criteria using pairwise comparisons

Once each criterion in the decision hierarchy has been defined and rated, the user must then weight the sub-criteria contributing to each "parent' criterion. CDA generates all possible combinations of each pair of criteria.

Each pair is compared using Saaty's "fundamental scale". When all pairs are compared, CDA then calculates the consistency ratio. If the ratio is greater than 0.1, the program alerts the user that the comparisons have resulted in an inconsistent evaluation. The user may then review and revise the comparisons, or ignore them.

4.3.5 Generating the results

Once criteria are defined, rated, and weighted, the user is then ready to generate the results. Three outputs can be generated from CDA:

- A map of five even classes produced from processing all the map overlays by reclassifying map field values to AHP ratings, multiplying each by the associated weight, and then summing the maps together for each level of the hierarchy. There is also a utility for automating the cartography by producing an ArcView layout and then writing the result out to a bitmap graphic file suitable for display on the Internet.
- A text document that records all criteria, the AHP ratings and weights for each criterion, and any descriptions and notes recorded during the process of building the decision hierarchy. This document is important because it lays a "paper trail" for the decisions that were made so they can be scrutinised by others and allow for refinement as data is improved or a better understanding of the problem is developed.
- An HTML file that contains the same information as the text document but suitable for display on the World Wide Web with hyperlinks to bitmap files produced by the automated cartography described above.

4.4 Biophysical Capability for Horticultural Crops in West Gippsland, Victoria

A key need in Victoria is regional assessments of agricultural potential for alternative crops. This is driven by a perceived need for diversification in agricultural production with the primary aim of increasing farm incomes. Few of these assessments have been done because of lack of soils data at an appropriate scale or because of lack of expertise within a specific region. These limitations still exist, though there is now a state-wide effort to improve natural resources data across the state and to maximise the knowledge of experts through published documents and "field days."

There is a program underway at present to develop 1:25,000 digital elevation models for the state (based on 10 metre contour intervals) and to refine soil maps that were generated at 1:100,000 to a scale accurate to 1:50,000. With these new datasets, it is possible to develop a wide range of new interpretations for agricultural and urban land use.

For agricultural capability mapping, techniques used to date have varied from study to study but generally have used the principle of "most limiting factor" or "single worst factor" (van de Graff 1988) to map constraints on agricultural productivity. This method has generally been applied to 1:250,000 land systems maps which are comprised of landscape units that combines attributes of topography, geology, vegetation, soils, and climate using a gestalt approach. The results of these evaluations have been of limited use because of the great degree of variation within each unit and the inability to replicate results because of the heavy reliance on expert judgement and lack of documentation of the process.

With the emphasis on improving techniques in land assessment, it was decided by State Government that it would be useful to trial the use of the AHP in assessment of biophysical capability analysis for horticultural crops. In this trial, four horticultural crops were selected, sweet corn, broccoli, carrots, and wine grapes. These three crops were selected because of their differences in soil and climate requirements and because of their current and potential economic value in the study area. All for crops were assessed using the AHP with the CDA, however, for brevity, only sweet corn will be discussed here.

4.4.1 Background to study area

The West Gippsland Catchment Management Authority Region (WGR) was legislatively defined by the Catchment and Land Protection Act 1994. The region lies in southeast Victoria, with its western boundary approximately 80 km east of Melbourne (Figure 6). It extends eastward to Lake Wellington, Dargo and Mt. Hotham and from the Great Dividing Range in the North to Bass Strait in the south. Major access is via the Princes and South Gippsland Highways.

The Maffra case study area is part of the Macalister Irrigation District (around the town of Maffra see Figure 6). This sub-region is selected because of the agricultural and horticultural advantages due to their proximity to supplementary water supplies for.

The WGR covers approximately 2,025,000 ha of which 1,009,190 ha (49%) is public land, most of which is forested. The remainder of the area is freehold with the predominant land uses of rain-fed dairy, beef and sheep grazing, and horticulture.

Figure 6. Location Map for West Gippsland study area in Victoria, Australia.

Most industries in West Gippsland draw directly on its natural resources. Agriculture, forestry, electricity, gas, and tourism are the most significant contributors to the regional economy. As well as those directly employed in the main industries, a significant number of people are employed in support industries, both in non-farm agricultural business and in other business, which are highly dependent on agriculture.

The total regional population is approximately 174,000 people. The ratio of urban to rural residents in the region is about 3:1 making it the most densely settled rural area in Victoria (Australian Bureau of Statistics 1998).

4.4.2 The AHP workshop

The application of the AHP for biophysical capability analysis for corn requires a panel of experts to define the decision hierarchy and criteria ratings and weightings. In this study Rob Dimsey, a horticulturalist with the Department of Natural Resources and Environment (DNRE) acted as the domain expert. In addition, John Williamson and Paul Rampant of the DNRE Centre for Land Protection Research (CLPR) acted as experts in land resource assessment and in a capacity as experts on the natural resource data for the study area. CLPR has produced the 1:25,000 digital elevation maps, the climate maps, and soils map in cooperation with Ian Sargeant, soils scientist and Mark Imhof of Agriculture Victoria. The general format of the workshop is as follows:

- A brief introduction to the AHP methodology is presented.
- Professionals from Centre for Land Protection Research then reviewed the fundamental databases and layers with the experts to familiarise them with the scale, accuracy and content of the GIS data.
- The definition of biophysical suitability is discussed, and questions about the process are answered.
- Using the CDA Software, the AHP hierarchy is developed and graphically built during the workshop; and comments, definitions of criteria, and other important information are recorded.
- Each criterion in the hierarchy is next discussed in detail. The map values are examined, ranked, and then rated on a 0 to 1 scale.
- Starting at the bottom of the hierarchy, criteria are then weighted against each other using the pairwise comparison technique as implemented in the CDA Software. The resulting weights for each criterion and the consistency ratios are reported immediately upon completion of the pairwise comparison. In the (rare) case where consistencies exceeded the recommended value of 0.1, the pairwise comparisons are reviewed and revised.

The resulting database was then taken to the Department of Geomatics at the University of Melbourne to process the data and generate each weighted map criterion. All the maps are divided into five equal classes. Resulting maps and a report of the criteria, definitions, weights, and ratings were then sent back to the original experts for review and revision.

The revised ratings were then made at the Department of Geomatics, and a new set of maps was finally generated. Table 1 provides an estimate of the time it took to perform the steps described above.

Table 1. Estimates of time to implement each step of the AHP.

Step	Estimated Time
1. Introduction to the AHP methodology	1/2 hour
2. Review of databases and layers	1/2 hour
3. Questions and answers	10 minutes
4. Construction of the AHP hierarchy	1 to 1 _ hours
5. Rating of map criterion	2 hours
6. Pairwise comparisons to weight criteria	1/2 to 1 hour
7. Map generation	4 hours
8. Review and revision	4 hours

The procedure as outlined above, proved to be effective. The expert participants were genuinely cooperative and took the procedure seriously. Rob Dimsey came prepared with reference books and maps to assist in the definition of criteria. The participation of the CLPR staff was critical in assisting in the interpretation of soil and climate attributes.

The use of the CDA software during the workshop was seen by participants to be a useful way of visualising progress as the decision tree was produced and as a helpful tool in keeping track of decisions that were made since any criteria could be reviewed simply by double clicking on its representation on the screen.

5. RESULTS

Figure 7 shows the decision hierarchy and weights for corn. Note that the weights for each level of the hierarchy sum to 1. Figure 8 shows the final AHP biophysical capability map for corn for the Maffra Study area generated by the CDA software. This is a composite of the soils map (Figure 9), slope map (Figure 10), and climate map (Figure 11).

6. CONCLUSIONS

The integration of the AHP with GIS provides a powerful tool for ranking sites based on multiple attributes. Software such as Catchment

Decision Assistant takes care of the enormously complex bookkeeping and analytical tasks required for such integration. By clear separation of the user interface from the analytical and database functions in a three tier software architecture, it is possible to use tools like CDA in a workshop environment to facilitate record keeping and provide a visual record of progress during the development of a decision hierarchy.

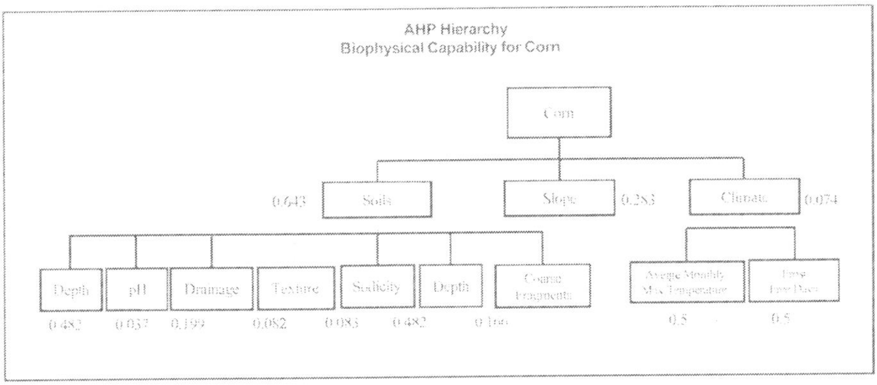

Figure 7. The AHP decision hierarchy for Corn

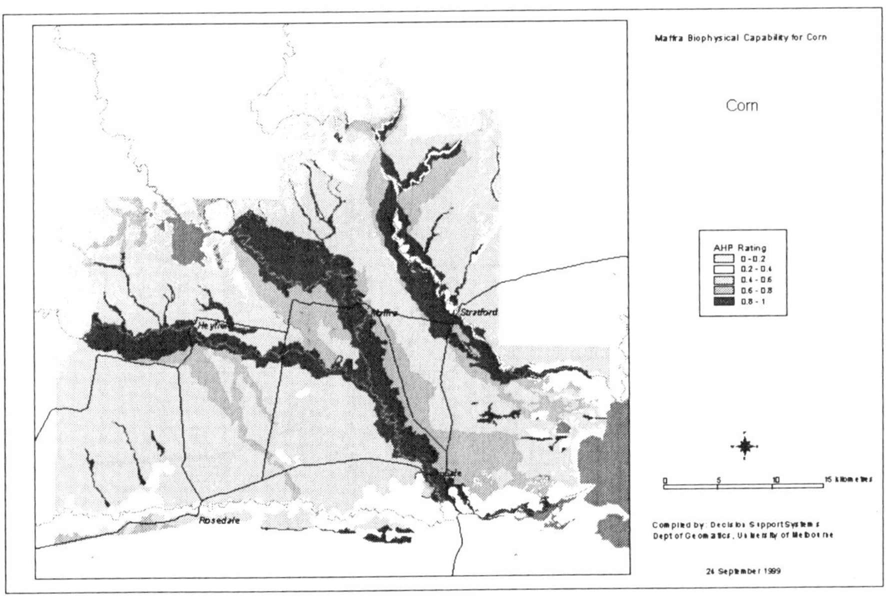

Figure 8. Final map showing biophysical capability site rankings for corn. Note the best areas are in low valleys along drainageways. These sites have a combination of good climate, gentle slopes, and good soil characteristics.

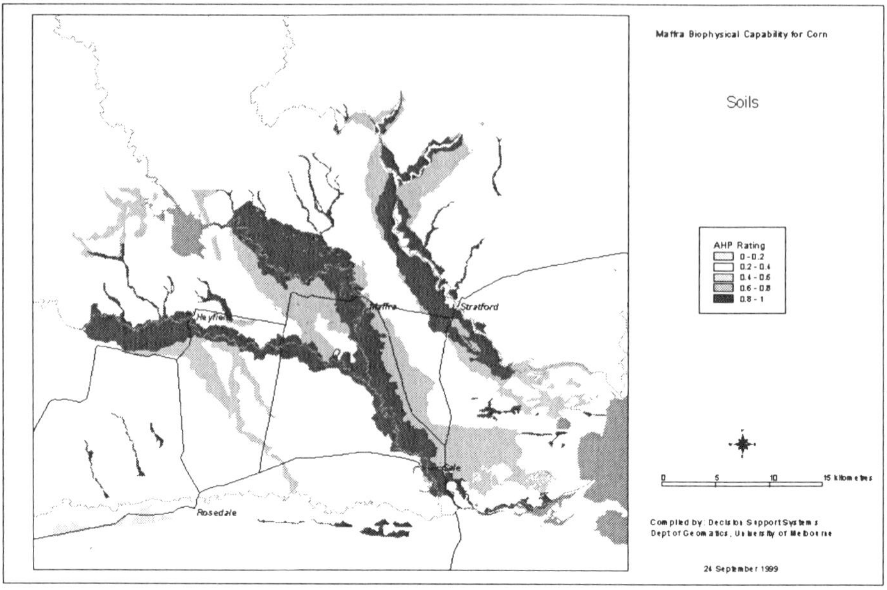

Figure 9. Biophysical capability for soils criteria. This map combines soil attributes for pH, surface and subsurface drainage, sodicity, soil texture, coarse fragments, and useable depth.

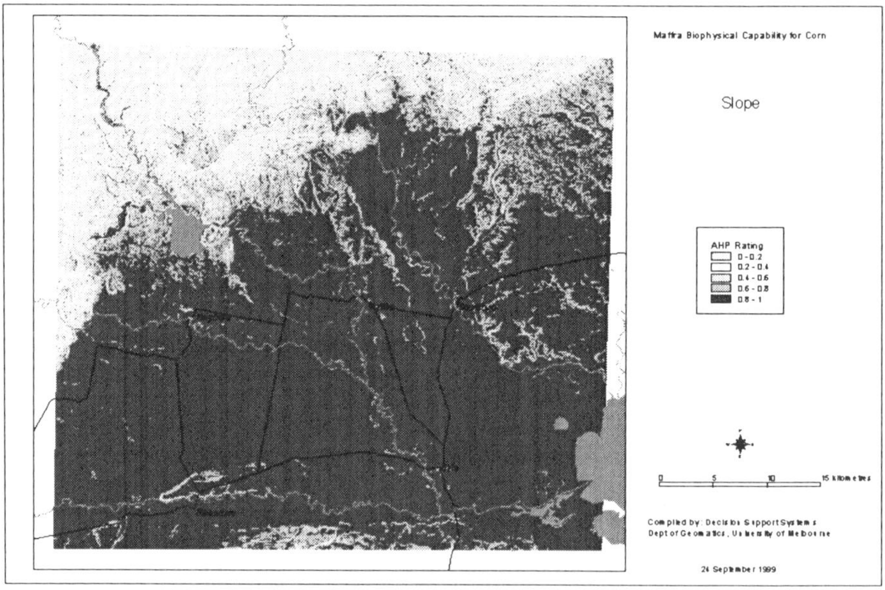

Figure 10. AHP ratings for slope map. The northern and northeastern regions of the study area are hilly and have slopes restrictive to cultivation.

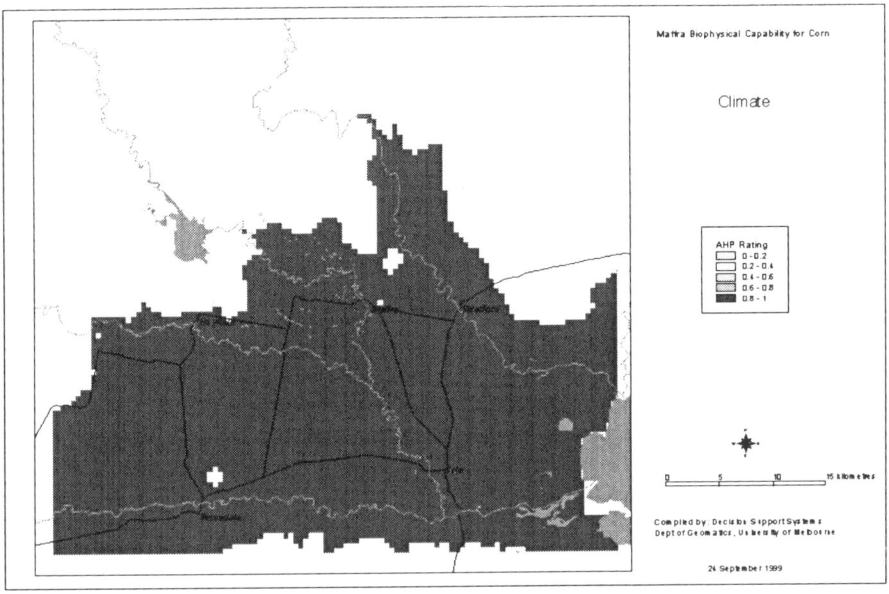

Figure 11. Climatic restrictions are also due primarily to topography with the hilly areas having shorter growing seasons and prone to frost.

The application of the AHP to biophysical land capability in West Gippsland using CDA software has shown the following benefits of the integration of GIS with the AHP:

- Interactive use of the software in workshop environments provides an interesting visual record for workshop participants. Because they can see comments recorded through the discussion, participants have confidence that their ideas are being recorded and are quick to clarify definitions that are not clear. The fact that comments and criteria are recorded on a computer offers an interesting psychological difference as compared to doing the same exercise on butcher paper. Participants find the process compelling and enjoy the participatory aspect of building the hierarchy on the computer screen. It is important however the person using the software in the workshop is proficient, and can type quickly to keep up with the flow of discussion. It is helpful to have an assistant record ideas on paper that may be lost during fast moving discussions.

- Integration of the AHP with GIS databases requires a map database with high levels of integrity. There can be no missing tables, or missing values. CDA uses a data management system developed for the DNRE decision support system that ensures the database is robust and accurate. CDA works successfully in this environment by supporting ArcInfo libraries, ArcView shape files and grid files. It is likely that any modelling software that automates access to large databases will need high quality data management of the sort underpinning CDA.

- Automation of the analytical and cartographic processes is essential in delivering high quality products, inexpensively and quickly. An essential characteristic of the AHP is the ability to re-examine results and quickly alter criteria or judgements. This type of feedback would be impossible in a GIS environment without the automation provided by software like CDA.
- It is clear the AHP can serve as a valuable tool in integrated catchment management by providing a decision making framework that captures the knowledge of experts, provides a systematic method for ranking alternatives, and with software like CDA creates a record so that decision criteria can be improved through time with better knowledge and data.

This has been a limited demonstration of the AHP in a specific case of resource assessment, however it does demonstrate the potential utility across a broad range of catchment issues.

ACKNOWLEDGEMENT

The authors would like to acknowledge the contribution of Peter Forbes (DNRE), John Williamson and Paul Rampant of the Centre for Land Protection Research in Agriculture Victoria, Ian Sargeant, Mark Imhof and Victor Sposito (Soil Chemistry Laboratory, Agriculture Victoria). The Bureau of Meteorology supplied the climatic data. Rob Dimsey, Senior Horticultural Industry Development Officer acted as expert in the development of the decision hierarchy. ArcInfo, ArcView and Spatial Analyst are trademarks of ESRI, Inc. Visual Basic and Access are trademarks of Microsoft Corporation. Funding for this project is provided by the Commonwealth of Australia through the Natural Heritage Trust and the Department of Natural Resources, State of Victoria.

REFERENCES

AUSTRALIAN BUREAU OF STATISTICS. 1998. CDATA96, 1996 Census of Population and Housing, Commonwealth of Australia.

BANAI-KASHANI, R. 1993. Fuzziness in geographical information systems: Contributions from the analytic hierarchy process, International Journal of Geographical Information, Vol. 7, No. 4, pp. 315-329.

BANAI-KASHANI, A.R. 1989. New method for site suitability analysis - the Analytic Hierarchy Process, Environmental Management, 13/6, 685-693.

BANTAYAN, N., AND BISHOP, I. 1998. Linking objective and subjective modelling for land use decision making. Landscape and Urban Planning, 43, pp. 35-48

GOLDEN, B.L., WASIL, E.A. AND LEVY, D.E. 1989. Applications of the analytic hierarchy process: a categorized, annotated bibliography, in: B.L. Golden, P.T. Harker and E.A. Wasil eds. Analytic Hierarchy Process, Springer-Verlag, New York, pp. 37-58.

ITAMI, B., MACLAREN, G. AND SPOSITO, V. 1999. The Victorian catchment management and sustainable agriculture decision support system. Proceedings of the MODSS '99 Conference, Brisbane, Australia, 1-6 August.

JANKOWSKI, P. AND RICHARD, L. 1994. "Integration of GIS-based suitability analysis and multicriteria evaluation in a spatial decision support system for route selection", Environment and Planning B: Planning and Design, Volume 21, pages 323-340.

SAATY,T. 1995. Decision Making for Leaders: The Analytical Hierarchy Process for Decisions in a Complex World, RWS Publications, Pittsburgh.

XIANG, X-N AND WHITLEY, D. 1994. Weighting land suitability factors by the PLUS method, Environment and Planning B; Planning and Design, Volume 21, pp. 273-304.

VAN DE GRAFF, R.H.M. 1988. Land Evaluation, Chapter 18, Australian Soil and Land Survey Handbook, edited by R.H. Gunn, J.A. Beattie, R.E. Reid and R.H.M. van de Graff. Inkata Press, Melbourne, Victoria, Australia.

Chapter 18

Past Developments and Future Directions for the AHP in Natural Resources

Daniel L. Schmoldt
USDA Forest Service, Southern Research Station, Madison WI USA

G. A. Mendoza
Department of Natural Resources and Environmental Sciences, University of Illinois, Urbana, Illinois USA

Jyrki Kangas
Finnish Forest Research Institute, Kannus, Finland

Key words: Social preferences, conflict resolution, consensus, participatory decision making.

Abstract: The analytic hierarchy process (AHP) possesses certain characteristics that make it a useful tool for natural resource decision making. The AHP's capabilities include: participatory decision making, problem structuring and alternative development, group facilitation, consensus building, fairness, qualitative and quantitative information, conflict resolution, decision support, and preferences structuring. For each of these facilities, we describe how it is reflected in land management and then illustrate how it is addressed by the AHP. Based on this analysis and on the preceding chapters of the book, we offer some suggestions for extending the AHP in new directions, e.g. peer-to-peer networking, site-specific management, forest management planning, statistical analyses, and software enhancements. The ability of the AHP to incorporate the human dimension (subjective preference) and to aid group decisions of choice are seen as the method's most noteworthy features.

D.L. Schmoldt et al. (eds.),
The Analytic Hierarchy Process in Natural Resource and Environmental Decision Making, 289–305.

1. INTRODUCTION

In chapter 1, we briefly outlined the nature of natural resource management in the context of ecosystem management—the current paradigm for land stewardship. Natural resource management, by and large, entails making choices among alternative courses of action, or more specifically, decisions about alternative management regimes. Making these decisions is problematic largely because of the decision environment's inherent complexity. Examples of these complications include: (1) multiplicity of management objectives, (2) involvement of several beneficiaries, or stakeholders, with their own demands (agendas) and concerns (belief systems), and (3) uncertainty emanating from a general lack of knowledge about the dynamic processes and relationships involving different ecosystem components. The argument presented earlier is that, in light of these underlying complexities, decision support tools are needed as instruments to make rational, carefully reasoned, and justifiable decisions in natural resource management.

The preceding chapters provide an overview of the analytic hierarchy process (AHP) and its broad application across a variety of natural resource and environmental problems. Those authors demonstrated the use of the AHP with other analytical tools (e.g., mathematical programming), for group and participatory decision making, as part of other decision methods (e.g., SWOT, SMART), and with extensions (e.g., fuzzy sets, GIS). In almost all chapters, a real-world example was also provided. While land management typically involves selecting among a relatively small set of possible alternatives, executing one of those alternatives is often irreversible and can have dramatic impacts. One of the general observations that should be taken away from those chapters is that even though the choice set is small, selecting the best one may be a very complex, and risky, decision. Yet, current decision methods often lack the necessary flexibility and sophistication to make a good choice and to support that choice later on.

This chapter has two general purposes. First, it briefly reviews some of the important functions of decision methods, particularly the AHP. This review, however, will put less emphasis on technical issues. The chapters contained in this book offer excellent expositions on both the technical aspects of the method, and the novel approaches used to apply the method to different problem situations. Second, based on this functionality analysis and on the innovative applications of, and extensions to, the AHP appearing in the contributed chapters of this text, we offer some suggestions for possible future directions for the AHP. We consider AHP enhancements as both new application options and as extensions to the AHP methodology itself.

2. AHP CAPABILITIES

Some of the desirable capabilities of the AHP have already been described, albeit obliquely, by the earlier chapters. The purpose of this section is to explicate and amplify those roles and to establish the enormous potential of the AHP. Hence, the presentation that follows describes these capabilities focusing on specific attributes that are compatible with distinctive characteristics of management issues in natural resources and the environment.

2.1 Participatory Decision Making

Natural resource management has become an arena for public involvement characterized by a dizzying array of stakeholder interests, both public and private. More and more, these interest groups demand a voice, both in policy making and management decisions. Increasingly, these groups have become more informed, better organized, assertive, and aggressive in their demands to be involved, not only as sources of information, but as active partners in decision making. For a natural resource management strategy to have any chance of success under these circumstances, it must adopt a genuine participatory approach, where each interest group has active involvement, with their voices heard and their input accommodated in the decision-making process.

Individual voting, or solicitation, of expert judgments via pair-wise comparisons is a feature of the AHP that is a good match for including multiple stakeholders. Each participant group can voice and record their own opinions in a hierarchy. Those voices can be treated equally or they can be weighted by importance, experience, prominence, or any other characteristic that distinguishes the individual groups. Furthermore, because a hierarchy is a recursive structure of sub-hierarchies, each group's judgments can become part of the overall decision process by affording each group their own sub-hierarchy. Within their sub-hierarchy, each group can formulate the decision problem in the way that makes the most sense to them. Because the overall hierarchy provides a record of participatory inclusion, it is readily apparent how stakeholders are incorporated into, and influence, the decision process. The explicitness of this process makes it much harder for groups to claim exclusion, "We weren't listened to," or for decision makers to falsely claim, "We included stakeholder input into our decision." The AHP doesn't force participatory decision making, but it facilitates it and records to what extent it was applied.

Several scenarios for conducting this multi-group process using the AHP were suggested in Schmoldt *et al.* (1995). These included: (1) each group formulates their own AHP decision hierarchy separately, (2) all groups together create a single hierarchy in a plenary session, or (3) each group creates a sub-hierarchy, which decision makers use as part of their overall decision hierarchy. In addition, groups' hierarchies can be pre-structured by top-level decision makers, with each group providing judgments only. Then, judgments can be obtained without face-to-face meetings, but by the use of mail surveys (q.v., Smith *et al.* 1995). By avoiding face-to-face meetings in this way, it is possible to mitigate many negative aspects of group dynamics. This last approach can be criticized for allowing decision makers to constrain stakeholder input, but it is still much better than allowing no input at all. These decision makers' overall hierarchy should still indicate how stakeholder input was eventually used in their final decision—which is the important thing.

In Finland, use of the AHP in participatory natural resource decision making has attracted a lot of attention, especially within the forestry sector. With state-owned forests in Finland covering one-third of all forest land, AHP principles have been widely applied in participatory strategic forest planning (Kangas 1999). However, the first participatory applications were carried out in nature conservation planning (Kangas 1994). The AHP has also been used in forest policy analysis at the province level (e.g., Kajala 1996). Recently, the AHP has mainly been used interactively in participatory decision support processes (Pykäläinen *et al.* 1999). Interactive use of the AHP has been found to be an effective teaching and learning tool that highlights the complexity of decision situations to participants and helps them understand existing trade-offs, as well as, competing interests. When integrated into the more general context of a participatory planning framework, an interactive AHP serves as a powerful means for successful conflict management.

2.2 Conflict Resolution

This is perhaps the most common issue in the natural resource management arena. Disagreements are most likely to arise among participants because of differences of opinions on substantive issues. Environmental problems, in particular, are traditionally delicate issues where deeply rooted beliefs and principles may stand in the way of achieving group consensus. Finding a responsible and perceptive way to resolve these differences or conflicts may ultimately determine the success or failure of management actions.

Saaty and Alexander (1989) describe some case studies showing the adaptability of the AHP for resolving conflicts, including political conflicts. In their text, different political conflicts were simulated using the AHP in order to understand conflicts better and to find ways to negotiate through them. The AHP was used as a tool to structure the different conflicts using their vital elements such as: the problem (level 1), parties in the conflict (level 2), objectives for each party (level 3), and basic political structures (level 4). Actions and judgments of the different actors were then simulated following a forward and backward process. The forward process is a generally descriptive process that identifies most likely outcomes given the influence of different parties. The backward process identifies desired outcomes and the necessary actions in terms of the hierarchy to achieve desired results. These case studies illustrate how the combination of these two processes applied in an AHP simulation environment can yield negotiable results.

Mendoza and Prabhu (2000) have also shown how a team of experts can be used to arrive at a collective decision with respect to assessing sustainability of forests. Inevitably, evaluating forest sustainability is a complex process, one that must involve experts from different disciplines. Due to the inherent complexity of the factors affecting sustainability, it is natural that assessments and professional views among experts also vary. In this study, the authors analysed different sets of indicators of forest sustainability proposed by the expert team. For some of these indicators, there were disagreements among experts as to their importance. Using the AHP, compromise sets were generated according to the relative weights of all indicators. The calculated relative weights served as objective measures by which indicators were prioritised. Hence, potential conflicts were avoided by using objective measures of relative importance that were calculated as a collective decision of all experts involved in the assessment.

2.3 Problem Structuring and Alternative Development

Many natural resource problems are shrouded with uncertainty because of a general lack of information or insufficient knowledge. Management objectives, for example, are not always known or, in some cases they are obscured and can only be elicited through prior analysis. Some aspects of the problem may also be undisclosed or not readily identifiable, although they may be articulated in qualitative terms. Hence, even before performing any analysis, problem conceptualisation and formulation need to be performed to gain a better understanding into the nature of a problem.

The decompositional and hierarchical features of the AHP offer a convenient platform for doing preliminary analysis. As shown in Chapter 1, the elements of a problem can be decomposed into manageable elements with decreasing levels of uncertainty or ambiguity. Decomposing a complex problem into a hierarchy of elements enables and conditions analysis where it is most appropriate. In the chapter by Mendoza and Prabhu (chapter 8), the problem of assessing forest sustainability illustrates hierarchy development. There, sustainability is decomposed into analytical constructs: from general principles to more tangible and measurable verifiers and parameters. Analyses were performed at each level independently but were linked and cumulated at higher levels in the hierarchy. In the chapter by Schmoldt and Peterson (chapter 7), fire modelling research issues are subdivided into *key questions*—and further into *responses* to those questions—within each of four research topic areas. Each topic area was assigned to a separate and independent workgroup, whose results were then aggregated by a research program manager at the highest level. Hence, decisions and judgments can be made at each level (or sub-hierarchy) of an AHP hierarchy, and finally, aggregated to produce impacts higher in the hierarchy.

SWOT analysis, a widely applied tool in strategic decision planning, offers one way to systematically approach a decision situation. However, SWOT provides no means to analytically determine the importance of factors or to assess the match between SWOT factors and decision alternatives. In, so called, A'WOT analysis (chapter 12), the AHP and its eigenvalue calculation framework are integrated with SWOT analysis. The AHP combined with SWOT yields analytically determined priorities for the factors included in SWOT analysis and makes them commensurable. In addition, decision alternatives can be evaluated with respect to each SWOT factor by applying the AHP (Pesonen *et al.* 2001). So, SWOT provides the basic frame within which to perform an analysis of the decision situation, and the AHP assists in carrying out SWOT analysis and in making more effective use of SWOT to develop alternative strategies and prioritise them.

In many cases, components of natural resource management problems are not known a priori; hence, they may have to be unveiled concurrently with analysis. The hierarchy offers a transparent framework where elements can be included or excluded interactively, and at any level in the hierarchy. Initially, decision makers may start with only a few elements (e.g., management options for a given objective). Then, with careful analysis, other elements may be added to progressively expand the scope of analysis. This is generally a better approach to complex natural resource problems, rather than starting too broad with limited knowledge of the elements or controllable actions. Iterative hierarchy development, analysis, and evaluation enable decision makers to create a dynamic decision process that can evolve over time and readily incorporates new information and knowledge as it becomes available.

2.4 Group Facilitation and Consensus Building

Because most natural resource management must take place in an environment conducive for public involvement and active participation, issues related to group dynamics, meeting facilitation, and consensus building have gained prominence (Schmoldt and Peterson 2000). Effective management has essentially become an exercise highly dependent on the ability to manage group interactions and to accommodate multiple inputs efficiently. The underlying goal is to manage or facilitate group interactions so that in the end some level of acceptable compromise can be achieved, unless consensus can be reached—the latter being a very rare event because of the diverse set of interests and concerns that characterize many natural resource problems.

The AHP, with its consistency measures, offers a pragmatic way to facilitate group decisions so that choices can be progressively and systematically steered toward an acceptable compromise. Consistency indices and consistency ratios can serve as guides to help direct the decision process towards better collective choices. The opportunity provided by the AHP for each participant to provide their input, and because these inputs are treated by the AHP in a manner transparent to the participants, it increases the likelihood that results of the analysis will be acceptable to all. This democratic process imparts ownership of any decision to the group as a whole.

The model described in Mendoza and Prabhu (1999) illustrates these points. In this model, experts were guided by the consistency index values to provide more consistent pair-wise comparisons of both the *indicators* and *verifiers* of sustainable forest management. Following an iterative process guided by the AHP's consistency indices, each expert (or forest sustainability assessor) was able to make more informed judgments leading to more consistent estimates of the relative importance of each sustainability indicator and verifier.

Kangas *et al.* (1998) used a traditional consensus building process, the Delphi technique, to quantify expert knowledge on forest biodiversity. To reduce bias, several independent experts carried out the required AHP pair-wise comparisons in a case study experiment. Variance components modelling was used to estimate judgment changes over three Delphi rounds for eleven experts. In this way, uncertainties in expert judgments elicited by pair-wise comparisons could be analytically studied, and the consistency of judgments could be improved during the process. It turned out that the judgments converged to some extent, while, in one case, an increase in shared inconsistency among judges was also detected. Variation between individuals decreased for all comparisons during the Delphi process.

Experiences by others (Peterson *et al.* 1994, Schmoldt *et al.* 1998), suggest that group participants seem to enjoy the search for consensus using the AHP and treat it somewhat like a game. Judgments offered by group members can be interleaved with feedback on group consistency—similar in some ways to the Delphi process noted above. There is no absolute requirement that consensus eventually arises, however; because, in the end, group judgments can be average to arrive at a group decision.

2.5 Fairness

The issue of fairness often surfaces in many group or participatory decision-making situations. The crux of the issue centres on the extent to which opinions of each participant are heard and considered as part of the decision process. In a democratic process, all opinions are weighted equally—one person, one vote. Realistically, however, some participants are more informed or are better positioned—either by skill, experience, or training—to provide better decisions. In such situation, the decision maker must decide whether to ascribe more importance to these "better" prepared participants, or to treat all participants equally regardless of expertise, experience, knowledge, or other extra-ordinary skills. The AHP is flexible enough to handle both situations. Because a "good" decision is an intellective choice and not a democratic (or majority or average) opinion, often it is preferable to treat individual opinions differentially. In this case, the AHP's aggregation procedure can assign different weights to each participant to reflect their varying degrees of expertise.

It should not necessarily be assumed, however, that knowledge in a field is coincident with analytical skill in that same field. Schmoldt and Peterson (2000) found that some group members, who were well respected and very knowledgeable in their field—and were instrumental in issue clarification and in AHP hierarchy development within their group—were, nevertheless, not as skilled at setting priorities (by making paired comparisons). It may be that the extensive knowledge possessed by those individuals enables them to see all sides of each issue so thoroughly that it clouds their ability to make critical comparisons and preferential choices. This suggests that *fairness* might best be achieved by allowing each participant to contribute in a way—which may not necessarily be voting or judging—that best utilizes their individual talents for the group's overall decision-making benefit.

2.6 Qualitative and Quantitative Variables

Informed decisions, whether they relate to common daily-life issues or to complex problems like natural resource management, rely on information which can be quantitative or qualitative. In general, better decisions are achieved not because of the abundance of data or information, but rather because of how well the information, qualitative or quantitative, is used. The AHP inherently uses mixed data. When quantitative data are available, and especially when the decision elements are not shrouded with ambiguity, pair-wise comparisons can become very precise. However, when quantitative data is inadequate, or in some cases nonexistent, participants may have to rely on intuition to make their judgements. These insights may be based on specialized experience or on general knowledge of known relationships among the decision elements.

Even in the presence of quantitative data, decision makers may wish to use subjective judgment to evaluate (or qualify) those numbers. Data-based numbers often imply a "counting" scale, which suggests that 100 of something is twice as good as (or twice as bad as, in other cases) 50 of the same thing. That sort of scaling may not necessarily reflect the inherent utility or value of those data, or the decision maker's preference. For example, the number of taxa present in a particular trophic level might be used to assess biodiversity—but 20 taxa present might, in reality, indicate that biodiversity is not much better than when 10 taxa are present. By using paired comparisons, the decision maker can create a preference scale for taxa counts. Similarly, one can also create mathematical relationships, e.g. using a logarithmic scale, but paired-comparison ratio scales are much easier for most decision makers to formulate and understand. In this same way, Saaty's chapter (chapter 2) describes how the 1-to-9 scale of the AHP can be extended to a 1-to-∞ scale, thereby expanding the realm of things that are commensurate.

2.7 Decision Support

Typically, one views "decision support" as data, as information, and as tools to manipulate and analyse those data. Decision support, however, can also include decision procedures that provide some measure of assurance that all pertinent issues and information have been fairly addressed in decision making. Public planning and the management of public lands are being subjected to increasing levels of scrutiny. Appeals and litigation often delay the implementation of management projects that were conceived with great effort and expense. The complexity of management issues and the reality of limited budgets, make it imperative that land management organizations have rational, consistent, and defensible management systems.

The AHP provides the structure and rigor to support complex and controversial decision making through its hierarchical framework and ratio-scale priority assignment. When examining an AHP hierarchy, it is immediately apparent how a decision was reached. While that does not preclude other decision makers from arriving at a different decision using a different hierarchy and different judgments, at least there is no doubt as to how the original decision was formulated. The AHP removes the mystery, and hidden rationale, from the decision process, so disagreements can focus on the real issues involved, and not on any inadequacies of the process itself.

2.8 Structuring Preferences

Accurate and complete information is critical to good decision making in natural resources management, not unlike other fields of endeavour. But, it is not the decision maker's only source for decision support. Knowledge, in the form of past experiences, (in)formal training, and beliefs/ideologies, all contribute to the process. This knowledge appears as a preference structure—a very selective lens, through which the decision maker views the world and interprets what he or she sees. One of the AHP's strengths is how it facilitate expression of those preferences—initially, as a set of comparison judgments and, ultimately, as priority vectors. Furthermore, preferences become even more evident *and* explicit because the final priority vector is a cardinal scale, rather than a less-informative ordinal scale. This also means that these priorities can be included in more quantitative analyses, such as mathematical programming, which are exampled in chapters 4-6, and in statistical tests for differences (Smith *et al.* 1995, Schmoldt *et al.* 1998). Use of paired comparisons seems to many to be a very natural and easy-to-understand method for stating preferences (Peterson *et al.* 1994), especially when compared to some other methods (Bard 1992). Preference structures elicited by the AHP aid in choice selection, are useful in subsequent analyses, and offer a glimpse into the belief systems that govern a decision maker's world view.

3. FUTURE DIRECTIONS AND EXTENSIONS OF THE AHP

The compatibility between AHP functionality and the general attributes of land management and decision making, as described in the above section, strongly intimates the AHP's potential as a decision support tool. This has also been borne out by the various applications described in the preceding chapters. The following subsections introduce some possible future extensions of the method to make it more appealing to a wider audience and their decision-making needs.

3.1 Site-Specific Decision Making

Advances in spatial, electronic, and digital technologies (precision forestry), particularly geographic information systems (GISs), are enabling land managers to formulate activities that address the unique needs of individual sites. GISs offer an environment within which the AHP can easily interface to make analyses of natural resource and environmental systems more site-specific. Itami *et al.* (chapter 17), for example, describes a computer-assisted decision support system combining GIS with the AHP. Similar efforts integrating the AHP with spatial analysis include Jankowski (1995), Jankowski *et al.* (1997), and Eastman *et al.* (1998). Making natural resource decisions site-specific adds realism and practicality to these decisions. Moreover, because of the AHP's flexible analytical features, it can take advantage of these spatial technologies and serve as a useful link to bridge information gaps using expert opinions (Store and Kangas 2001). Strengthening this link will mutually enhance the applicability of the AHP as well as the utility of these spatial tools, which ultimately should enhance the acceptability and practicality of natural resource use decisions.

3.2 Peer-to-Peer Networking

More and more land management decisions are being made in a group context, which may include a broad spectrum of resource specialists or a diverse set of stakeholder organizations. In either case, there are logistic difficulties in organizing such group meetings around everyone's busy schedule, so that everyone is coincident in both space and time. Tele- and video-conferencing can address the spatial differences, but not the time differences. Everyone must still be available at an appointed time to participate in a conferencing call.

An emerging new networking paradigm, peer-to-peer, is gaining popularity with certain applications, e.g., the sharing of computer processing time over the Internet to solve highly computational problems. This differs dramatically from the client-server protocols that we have become familiar with using the Internet, e.g., FTP, POP3, HTTP. In peer-to-peer networking there is a direct interchange of information between computers at many different locations, without any distinction between one computer providing services and one receiving those services. By combining peer-to-peer networking with AHP software designed to operate in this environment, decision makers working in different *locations* could contribute to an AHP decision process at different *times*. In such a scenario, several AHP decision hierarchies might be created and exist simultaneously, or there might be a single one that everyone is working on together. This type of computer-mediated work environment has been promoted in the literature on group-supported cooperative work (Engelbart and Lehtman 1988), wherein computerized documents and tools provide the foci and capabilities for multiple participants to author a common document collaboratively. A Java version of the AHP (Schmoldt and Lu, unpublished) already exists that runs on all computer platforms. There are plans to add a networking component, which would allow this type of distributed group decision making and relieve participants of the time and space constraints associated with most traditional group activities.

3.3 Extending and Embedding AHP Software

The immediately preceding section emphasized combining the AHP with a GIS. In fact, both chapters 16 and 17 describe using the AHP with such spatial tools. The opportunity also exists for *embedding* the AHP in other software tools. Because the AHP can be used to describe and analyse behavioural decision making, it can be viewed as a useful knowledge acquisition tool (Schmoldt 1998). It could be included as one of a suite of tools that aids the interview process. There are also many forest/ecosystem management software tools (e.g., NED, Twery *et al.* 2000) that could benefit from goal priority setting. For most land managers, all goals do not carry equal importance, so our management aids need to accommodate those preferences. AHP software itself can also be *extended* by some of the priority analysis methods proposed in chapter 15 (see below) and by the inclusion of uncertainty using SMART (chapter 13). While many software implementations of the AHP include some sensitivity analysis capability, they are quite limited. The use of TreeMaps (Asahi *et al.* 1995)—a multi-level analysis tool—significantly enhances a decision makers investigation of "what-if" scenarios. The Java version of the AHP mentioned above includes this TreeMaps feature. As useful as the AHP method is by itself, it is even more valuable when merged with other software or when its own implementations are extended in meaningful ways.

3.4 Forest Management Planning

Although the AHP has achieved good success in strategic natural resources planning, some problems have also been noticed. One drawback is that when "alternatives" represent composite actions, scoring each alternative (even using absolute rating) can easily become a complex task. For example, in forest management planning—within an area consisting of possibly hundreds of forest stands each having several alternative treatment schedules—there are far too many possible forest plan alternatives (i.e., combinations of stand-wise schedules) to be evaluated and compared. In that kind of situation, the AHP alone is not enough; efficient optimisation algorithms are needed to analyse production possibilities, and to produce alternative strategies and compare them.

One possibility is to utilise a hybrid approach, combining the AHP and other decision support techniques. In a successful hybrid approach, shortcomings of one method can be mitigated by utilising the benefits of other methods. The HERO heuristic optimisation method is an example of a practical hybrid that makes use of the AHP and numerical optimisation (chapter 4). It is specially developed for tactical forest planning to analyse a great number of alternative management plans. One of the key ideas in HERO is to utilize principles of the AHP in the formulation of the optimisation problem in a manner more compliant with the objectives and preferences of the decision maker than is possible using mathematical programming alone. In addition, combined use of the AHP and goal programming has been proposed for similar purposes (chapter 6). Integrating the AHP into more process-oriented approaches, having their foundations in general decision theories has been found a promising approach for participatory decision-making processes. The combination of the AHP and Positional Analysis (chapter 9) is an example of hybrid methods usable in participatory planning. The hybrid method A'WOT (chapter 12) also represents an approach where the AHP is applied within a more general strategic management framework (SWOT).

There is still plenty room for advances in this area of methodological development work. Most likely, one of the main directions for AHP-related research in the future will focus on integrating ideas, techniques, and methods appearing in other theories of decision support.

3.5 Statistical Analysis within the AHP Framework

The lack of a sound statistical theory behind the AHP has also been seen as one of the drawbacks of the method (e.g., Alho *et al.* 1996). In practical applications, too, problems have arisen regarding use of the standard AHP, that can be alleviated by application of statistical methods. Perhaps, the two foremost problems in this sense are that the original comparison scale does not allow expression of any hesitation regarding comparisons, and that the AHP itself does not provide tools for a thorough analyses of the priorities, particularly the uncertainty inherent in the data. However, the basic idea of performing pairwise comparisons, as being a pedagogical and intuitive approach, has proved to be very practicable.

Already in the 1980's, de Jong (1984) and Crawford and Williams (1985) showed how pairwise comparison data could be analysed by using a regression model instead of the eigenvalue technique. In many cases, the two methods give similar numerical results, but one major difference is that the regression model enables an analysis of uncertainties. As an extension to the work of de Jong, Crawford and Williams (1985), and Alho *et al.* 1986), Alho and Kangas (1997) proposed a Bayesian approach to the regression model, which provides results that may be more easily understood by decision makers than p-values from classical hypothesis tests. Leskinen and Kangas (1998), in turn, showed how to analyse interval judgment data—instead of judgments given as a single number—in the Bayesian regression framework. Furthermore, Alho *et al.* (chapter 15) showed how the characteristics of the attributes being compared, or the background characteristics of the judges, could be incorporated into the regression model as explanatory variables. In Chapter 15, they also illustrated how the regression approach permits estimation of priorities based on fewer pairwise comparisons. This allows one to consider more decision elements than the standard AHP.

Using statistical analyses does not violate any principles of the AHP. Instead, they serve as additional tools for decision support carried out within the AHP framework. As such, they provide decision makers with greater information regarding their preferences and choices.

4. CONCLUSIONS

Technological advances continue to increase rapidly. Most notably, these are arriving in the form of new and innovative decision support tools. Similarly, improvements in data generation, storage, processing, and management are reducing information gaps and data needs. Finally, we are also realizing transformations to methodologies that address the human dimensions of resource management. This is the area within which the AHP fits, as it puts the decision maker at centre stage and allows him/her to effectively utilize the volumes of information generated by the other technologies. It provides a mechanism to organize and condense information so that it can articulate a choice in the mind's eye of the decision maker.

In looking back at the many examples provided in the text, there are a surprising number that deal with decision making in a group setting. How readily the AHP accommodates group processes strongly argues for its use in a wide variety of applications. This is reflected in its value for participatory activities, fairness concerns, consensus building, and conflict resolution. The interdisciplinary nature of resource management issues and recent stakeholder inclusion in the decision process makes those AHP features most compelling.

As the AHP becomes as widely known as other multi-objective decision methods, it should gain more prominence in natural resource management applications. Decision makers, that we have introduced to the method, are very pleased with it and agree that it is a very useful tool. However, all such decision processes enjoy limited use in practice, seemingly for other reasons. While new analytical tools, e.g. GISs, and innovative data collection/storage methods are readily adopted by land management organizations, techniques for actually making decisions—choosing alternatives—are less easily accepted or used. Because the act of making a decision is inherently risky and error prone, many managers avoid the decision process or, at least, do not want the process laid open to examination and possible criticism. Consequently, the steps and rationale actually used in making choices are often confusing and shrouded in mystery. As noted elsewhere in the text, it then becomes difficult to justify decisions when they are scrutinized, which opens the door to contentious arguments and possible litigation. Therefore, what hinders the AHP's use most (and other decision methods, also) may be established procedures and protocols and institutional inertia, rather than any failings of the method's approach. By highlighting this final step of land management decision making (i.e., choice), we hope to encourage more regular and committed use of the available methods.

REFERENCES

ALHO, J.M. AND KANGAS, J. 1997. Analyzing uncertainties in experts' opinions of forest plan performance. Forest Science 43: 521-528.

ALHO, J.M., KANGAS, J. AND KOLEHMAINEN, O. 1996. Uncertainty in the expert predictions of the ecological consequences of forest plans. Applied Statistics 45: 1-14.

ASAHI, T., TURO, D. AND SHNEIDERMAN, B. 1995. Using Treemaps to visualize the analytic hierarchy process. Informations Systems 6: 357-375.

BARD, J. F. 1992. A comparison of the analytic hierarchy process with multiattribute utility theory: A case study. IIE Transactions 24: 111-121.

CRAWFORD, G. AND WILLIAMS, C. 1985. A note on the analysis of subjective judgment matrices. Journal of Mathematical Psychology 29: 387-405.

DE JONG, P. 1984. A statistical approach to Saaty's scaling method for priorities. Journal of Mathematical Psychology 28: 467-478.

EASTMAN, J. R., JIANG, H., AND TOLEDANO, J. 1998. Multi-criteria and multi-objective decision making for land allocation using GIS. Pages 227-252 in Beinat, E. and Nijkamp, P. (eds.) Multicriteria Analysis for Land-Use Management. Kluwer Academic Publishers, Dordrecht.

ENGELBART, D. AND LEHTMAN, H.. 1988. Working smarter. BYTE 13: 245-252.

JANKOWSKI, P. 1995. Integrating geographical information systems and multiple criteria decision-making methods. International Journal of Geographical Information Systems 9: 251-273.

JANKOWSKI, P., NYERGES, T. L., SMITH, A., MOORE, T. J., AND HORVATH. E. 1997. Spatial group choice a SDSS tool for collaborative spatial decision making. International Journal of Geographical Information Systems 11: 576-602.

KAJALA, L. (ed.). 1996. Lapin metsästrategia. [Strategy for the forest of Finnish Lapland, In Finnish]. Maa- ja metsätalousministeriön julkaisuja 2/1996.

KANGAS, J. 1994. An approach to public participation in strategic forest management planning. Forest Ecology and Management 70: 75-88.

KANGAS, J. 1999. The analytic hierarchy process (AHP): Standard version, forestry application and advances. Pages 96-105 in Helles, F., Holten-Andersen, P. and Wichmann, L. (eds.) Multiple Use of Forests and Other Natural Resources. Kluwer Academic Publishers, Amsterdam.

KANGAS, J., ALHO, J., KOLEHMAINEN, O. AND MONONEN, A. 1998. Analysing consistency of experts' judgments - Case of forest biodiversity. Forest Science 44: 610-617.

LESKINEN, P. AND KANGAS, J. 1998. Analysing uncertainties of interval judgment data in multiple criteria evaluation of forest plans. Silva Fennica 32: 363-372.

MENDOZA, G. A. AND PRABHU, R. 2000. Multiple criteria analysis for assessing criteria and indicators in sustainable forest management: A case study on participatory decision making in a Kalimantan forest. Environmental Management 26: 659-673.

PESONEN, M., KURTTILA, M., KANGAS, J., KAJANUS, M. AND HEINONEN, P. 2001. Assessing the priorities among resource management strategies at the Finnish Forest and Park Service. Forest Science (in press).

PETERSON, D. L., SCHMOLDT, D. L. AND SILSBEE, D. G. 1994. A case study of resource management planning with multiple objectives and projects. Environmental Management 18: 729-742.

PYKÄLÄINEN, J., KANGAS, J. AND LOIKKANEN, T. 1999. Interactive decision analysis in participatory strategic forest planning: Experiences from state-owned boreal forests. Journal of Forest Economics 5: 341-364.

SAATY, T. L. AND ALEXANDER, J. M. 1989. Conflict Resolution: The Analytic Hierarchy Process. Praeger Publishers, New York.

SCHMOLDT, D. L. 1998. Knowledge acquisition using linguistic-based knowledge analysis. AI Applications 12: 1-20.

SCHMOLDT, D. L. AND PETERSON, D. L. 2000. Analytical group decision making in natural resources: Methodology and application. Forest Science 46: 62-75.

SCHMOLDT, D. L., PETERSON, D. L., KEANE, R. E., LENIHAN, J., MCKENZIE, D. M., WEISE, D. R. AND SANDBERG, D. V. 1998. Modeling the effects of fire disturbance on ecosystsm: A scientific agenda for research and management. USDA For. Serv. Gen. Tech. Report PNW-GTR-455. Pacific Northwest Research Station, Portland OR

SCHMOLDT, D. L., PETERSON, D. L. AND SMITH, R. L. 1995: The analytic hierarchy process and participatory decisionmaking. Pages 129-143 In Power, J. M., Strome, M. and Daniel, T. C. (eds.), Proceedings of the 4th International Symposium on Advanced Technology in Natural Resources Management. American Society of Photogrammetry and Remote Sensing, Bethesda MD

SMITH, R. L., BUSH, R. J. AND SCHMOLDT, D. L. 1995. A hierarchical analysis of bridge decision makers. Wood and Fiber Science 27: 225-238.

STORE, R. AND KANGAS, J. 2001. Integrating spatial multi-criteria evaluation and expert knowledge for GIS based habitat suitability modelling. Landscape Planning (in press).

TWERY, M. J., RAUSCHER, H. M., BENNETT, D. J., THOMASMA, S. A., STOUT, S. L., PALMER, J. F., HOFFMAN, R. E., DECALESTA, S. A., GUSTAFSON, E., CLEVELAND, H., GROVE, J. M., NUTE, D., KIM, G. AND KOLLASCH, R. P. 2000. NED-1: Integrated analysis for forest stewardship decisions. Computers and Electronics in Agriculture 27: 167-193.

Managing Forest Ecosystems

1. K. von Gadow, T. Pukkala and M. Tomé (eds.): *Sustainable Forest Management.* 2000 ISBN 0-7923-6356-6
2. K. von Gadow (ed.): *Risk Analysis in Forest Management.* 2001 ISBN 0-7923-6900-9
3. D.L. Schmoldt, J. Kangas, G.A. Mendoza and M. Pesonen (eds.): *The Analytic Hierarchy Process in Natural Resource and Environmental Decision Making.* 2001 ISBN -7923-7076-7

Kluwer Academic Publishers – Dordrecht / Boston / London